Advances in Online Chemistry Education

ACS SYMPOSIUM SERIES **1389**

Advances in Online Chemistry Education

Elizabeth Pearsall, Editor
York Technical College
Rock Hill, South Carolina, United States

Kristi Mock, Editor
University of Toledo
Toledo, Ohio, United States

Matt Morgan, Editor
Western Governors University
Salt Lake City, Utah, United States

Brenna A. Tucker, Editor
University of Alabama at Birmingham
Birmingham, Alabama, United States

Sponsored by the
ACS Division of Chemical Education

American Chemical Society, Washington, DC

Library of Congress Cataloging-in-Publication Data

Names: Pearsall, Elizabeth, editor.
Title: Advances in online chemistry education / Elizabeth Pearsall, Editor, York Technical College, Rock Hill, South Carolina, United States, Kristi Mock, Editor, University of Toledo, Toledo, Ohio, United States, Matt Morgan, Editor, Western Governors University, Salt Lake City, Utah, United States, Brenna A. Tucker, Editor, University of Alabama at Birmingham, Birmingham, Alabama, United States ; sponsored by the ACS Division of Chemical Education.
Description: Washington, DC : American Chemical Society, [2021] | Series: ACS symposium series ; 1389 | Includes bibliographical references and index.
Identifiers: LCCN 2021039708 (print) | LCCN 2021039709 (ebook) | ISBN 9780841298231 (hardcover) | ISBN 9780841298224 (ebook other)
Subjects: LCSH: Chemistry--Study and teaching. | Chemistry--Computer-assisted instruction. | Chemistry--Web-based instruction.
Classification: LCC QD49.5 .A38 2021 (print) | LCC QD49.5 (ebook) | DDC 540.71--dc23
LC record available at https://lccn.loc.gov/2021039708
LC ebook record available at https://lccn.loc.gov/2021039709

The paper used in this publication meets the minimum requirements of American National Standard for Information Sciences—Permanence of Paper for Printed Library Materials, ANSI Z39.48n1984.

Copyright © 2021 American Chemical Society

All Rights Reserved. Reprographic copying beyond that permitted by Sections 107 or 108 of the U.S. Copyright Act is allowed for internal use only, provided that a per-chapter fee of $40.25 plus $0.75 per page is paid to the Copyright Clearance Center, Inc., 222 Rosewood Drive, Danvers, MA 01923, USA. Republication or reproduction for sale of pages in this book is permitted only under license from ACS. Direct these and other permission requests to ACS Copyright Office, Publications Division, 1155 16th Street, N.W., Washington, DC 20036.

The citation of trade names and/or names of manufacturers in this publication is not to be construed as an endorsement or as approval by ACS of the commercial products or services referenced herein; nor should the mere reference herein to any drawing, specification, chemical process, or other data be regarded as a license or as a conveyance of any right or permission to the holder, reader, or any other person or corporation, to manufacture, reproduce, use, or sell any patented invention or copyrighted work that may in any way be related thereto. Registered names, trademarks, etc., used in this publication, even without specific indication thereof, are not to be considered unprotected by law.

Foreword

The ACS Symposium Series is an established program that publishes high-quality volumes of thematic manuscripts. For over 40 years, the ACS Symposium Series has been delivering essential research from world leading scientists, including 36 Chemistry Nobel Laureates, to audiences spanning disciplines and applications.

Books are developed from successful symposia sponsored by the ACS or other organizations. Topics span the entirety of chemistry, including applications, basic research, and interdisciplinary reviews.

Before agreeing to publish a book, prospective editors submit a proposal, including a table of contents. The proposal is reviewed for originality, coverage, and interest to the audience. Some manuscripts may be excluded to better focus the book; others may be added to aid comprehensiveness. All chapters are peer reviewed prior to final acceptance or rejection.

As a rule, only original research papers and original review papers are included in the volumes. Verbatim reproductions of previous published papers are not accepted.

ACS Books

Contents

Preface ix

1. **Promoting Student Learning and Engagement: Data-Supported Strategies from an Asynchronous Course for Nonmajors** 1
   Laura E. Simon, Marcia L. O. Kloepper, Laurel E. Genova, and Kathryn D. Kloepper

2. **Transitioning from High-Stakes to Low-Stakes Assessment for Online Courses** 21
   Matthew D. Casselman

3. **A First Semester General Chemistry Flipped Remote Classroom: Advantages and Disadvantages** 35
   Wendy E. Schatzberg

4. **Problem Based Learning Group Projects in an Online Format – A Sequential Approach** 45
   Simona Marincean and Marilee A. Benore

5. **Maintaining Rigor in Online Chemistry Courses - Lessons Learned** 59
   Mitzy Erdmann, Sithira Ratnayaka, Brenna A. Tucker, and Elizabeth Pearsall

6. **A How-To Guide for Making Online Pre-laboratory Lightboard Videos** 77
   Timothy R. Corkish, Max L. Davidson, Christian T. Haakansson, Ryan E. Lopez, Peter D. Watson, and Dino Spagnoli

7. **Working It Out: Adapting Group-Based Problem Solving to the Online Environment** ... 93
   J. L. Kiappes and Sarah F. Jenkinson

8. **Teaching Chemistry Down Under in an "Upside Down" World: Lessons Learned and Stakeholder Perspectives** 105
   Elizabeth Yuriev, Andrew J. Clulow, and Jennifer L. Short

9. **Student Experiences and Perceptions of Emergency Remote Teaching** 123
   Barbara Chiu and Nicole Lapeyrouse

10. **Students as Partners: Co-creation of Online Learning to Deliver High Quality, Personalized Content** 135
    Amy L. Curtin and Julia P. Sarju

11. **Options and Experiences for Online Chemistry Laboratory Instruction** 165
    Matt Morgan and Emily Faulconer

12. Lessons Learned from Implementing Blended and Online Undergraduate Chemistry Laboratory Teaching during the Covid-19 Pandemic 177
Helen Cramman, Mia A. B. Connor, Chapman Hau, and Jacquie Robson

Editors' Biographies 195

Indexes

Author Index 199

Subject Index 201

Preface

In 2019, a small group of chemistry educators submitted symposium sessions and workshops for the 2020 Biennial Conference on Chemical Education (BCCE) conference related to teaching chemistry lecture and laboratory in the online environment. After the subsequent cancelling of the BCCE conference due to the COVID-19 pandemic, the idea for this eBook was formed. The evolution of this eBook reflects what we do as scientists and educators when new information is presented – incorporate the information and make appropriate adjustments. With that in mind, this eBook is reflective of topics that were planned for the 2020 BCCE conference combined with lessons learned from wide-scale switch to remote and online instruction as a result of the COVID pandemic. By the nature of these events, this eBook builds upon the COVID eBook special. The lessons learned from the transition to emergency remote courses provide necessary perspectives and information for those designing an online chemistry course in the future.

Over the years, many terms have emerged to describe distance learning - Hi-Flex, remote, online, virtual, and more. For the purposes of this eBook and preface, *emergency remote* instruction will refer to courses that were moved to an online environment as a result of the COVID-19; whereas, *online* courses refer to those which are designed and developed specifically for asynchronous or synchronous delivery in an online environment.

Online education has grown consistently in the last fifteen years in the United States. According to the National Center for Education Statistics (NCES), more than 6.9 million undergraduate and post baccalaureate students were enrolled in at least one online course in Fall 2018. The data also show that from 2012 to 2016, the number of students not taking any distance education courses declined by 1.7 million students. Thus, to continue to meet the needs of students, educators must increase the availability of online courses. The COVID-19 pandemic moved these efforts forward almost overnight. The duration of the pandemic has largely provided an opportunity for faculty to design and develop online chemistry courses for the long-term. However, many chemical educators lack a background in pedagogy and instructional methodologies for both in-person and online courses. As a result, chemical educators need to seek out and share pedagogical knowledge, online instructional design tips, and best practices in online chemistry education for the online chemistry classroom to ensure that the online course that are being offered are of the highest quality.

The importance of ensuring the design of a chemistry course for the appropriate level of student cannot be emphasized enough in the online classroom. Activities and assignments that are too hard will frustrate students while those that are too easy may result in dubbing assignments as "busy work" and consequently cause students neglect necessary learning activities. While these adaptations are necessary, they are often challenging and require the exploration of instructional methods and technologies that are frequently unused in chemistry education. The authors of chapters 2 - 6 focus on essential topics related to online pedagogy: ensuring regular and substantive interaction (2), adapting assessments for an online organic chemistry classroom (3), teaching chemistry using a flipped classroom online (4), problem based group projects in introductory, gateway, and upper division courses (5), and methods for maintaining rigor in an online chemistry classroom (6). Chapter 7 delves into using lightboard videos in the online classroom for pre-laboratory content and chapter 8 introduces peer-to-peer workshops for problem-oriented sessions in the online classroom. While many chapters herein share student perspectives about specific components of an online course, chapters 9 and 10 do this on a larger scale. These two chapters dive into the decision making process for designing online chemistry courses (9) and share student data about the emergency remote transition due to the pandemic (10). In a unique spin on course design, a collaborative model for online chemistry course development is shared in chapter 11 where students co-created the content in online chemistry courses at the University of York.

Historically, chemistry laboratory courses are conducted in-person, in a laboratory environment, regardless of the delivery method of the lecture content. Due to the circumstances surrounding the COVID-19 pandemic, virtually all faculty teaching on-ground were required to adjust, edit, and develop new methods for laboratory instruction almost overnight. Chemistry faculty employed a variety of instructional tactics to ensure students were continuously able to learn during the immediate transition in Spring 2020 with ongoing adjustments necessary in subsequent terms. The last two chapters of this book explores different aspects of teaching a chemistry laboratory in an online environment. Chapter 12 explores the myriad of options related to teaching a chemistry laboratory online including simulations, kitchen chemistry, and physical lab kits while chapter 13 shares compares student experiences from a fully online lab compared to a blended model.

As evidenced in this eBook, the COVID-19 pandemic forced many institutions and faculty to quickly adopt and expand their online teaching, or necessitated much shorter timelines for planned initiatives related to teaching chemistry online. Historically, catastrophic events have compelled advancement in technology that has provided long-term benefit, as seen with the development of the radar during World War II. In the same way, lessons learned from the pandemic can be applied to online chemistry education. Documents like this eBook provide a record of pedagogical models, novel methodologies, student feedback, and perspectives of teaching chemistry online, which can be leveraged in the future exploration and expansion of chemistry courses taught in the online environment.

Elizabeth Pearsall
Associate Dean
Institute for Teaching Excellence
York Technical College
452 S. Anderson Rd.
Rock Hill, SC 29730, United States

Kristi Mock
Associate Lecturer
Department of Chemistry and Biochemistry
University of Toledo
2801 W. Bancroft, MS 602
Toledo, OH 43606, United States

Matt Morgan
Western Governors University
4001 S 700 E #300
Salt Lake City, UT 84107, United States

Brenna A. Tucker
Introductory Chemistry Coordinator
Department of Chemistry
University of Alabama at Birmingham
901 14th Street South
Birmingham, AL 35294, United States

Chapter 1

Promoting Student Learning and Engagement: Data-Supported Strategies from an Asynchronous Course for Nonmajors

Laura E. Simon,[1] Marcia L. O. Kloepper,[2] Laurel E. Genova,[3] and Kathryn D. Kloepper[1,*]

[1]Department of Sociology, Mercer University, 1501 Mercer University Drive, Macon, Georgia 31207, United States
[2]Center for Digital Learning, University of New Mexico, Albuquerque, New Mexico 87131, United States
[3]Tift College of Education, Mercer University, 1501 Mercer University Drive, Macon, Georgia 31207, United States
[4]Department of Chemistry, Mercer University, 1501 Mercer University Drive, Macon, Georgia 31207, United States
*Email: kloepper_kd@mercer.edu

This chapter addresses common concerns related to teaching online courses with laboratory components. The concerns align with three types of interactions central to student learning and engagement: student interactions with their instructor, student interactions with course content, and student interactions with their peers. Drawing on observations from instructors, strategies from an instructional designer, and student data from an online, asynchronous chemistry lab course for nonmajors, this chapter provides recommendations that promote student engagement and learning in online chemistry courses. These findings and recommendations may translate to other online and hybrid science courses. Specific suggestions for transitions back to in-person classes also are highlighted.

Introduction

The COVID-19 pandemic required fully face-to-face institutions to consider how best to pivot in-person instruction to remote delivery, and many chemistry instructors were tasked with the additional challenge of moving lab experiences online (*1*, *2*). However, few would argue that this rapid switch in March 2020 is equivalent to a fully online course; while courses delivered online share outcomes and pedagogical approaches with their face-to-face counterparts, each requires deliberate planning that can be specific to delivery mode (*3*, *4*).

With many institutions still delivering instruction online as of the time of this writing, it remains to be seen what long-term online offerings will persist after institutions go back to full-time in-person instruction as well as what effects, if any, the "temporary" online experience has on future chemistry instruction (2). Even with prior remote instruction experience, designing an online chemistry course can be a daunting process. Adding in online lab requirements can heighten instructor concerns about learning and engagement, especially given the gaps in understanding of how students learn in a laboratory setting (5–12).

This chapter addresses common concerns related to fostering student learning and engagement in online lab-based chemistry courses (Table 1). These concerns fall into three learner interaction categories: with their instructor, with course content, and with their peers. The concerns in Table 1 are especially relevant and common for instructors designing or adapting a course for online delivery for the first time, when they may not be as experienced with online pedagogies. The core themes from Table 1 align with some features of the community of inquiry framework, which outlines three key areas of presence for online learning: teaching presence, cognitive presence, and social presence (13, 14). Suggested resources to learn more about this framework and how to develop each presence in an online class include the Community of Inquiry website and the Center for Teaching Excellence of the University of Virginia (15, 16).

Table 1. Common concerns about online courses.

Student-Instructor Interaction Concerns	Student-Content Interaction Concerns	Student-Peer Interaction Concerns
• What should the first day of class look like online? • What is the best strategy for communicating with students in an online class? • How can student-instructor interactions be fostered in online settings?	• How will students stay engaged with course material, especially in asynchronous classes? • What is the appropriate balance of course work? • What online assessments are best at promoting student learning? How will lab skills be assessed? Will students learn if they aren't proctored?	• How can student-student interactions be fostered in online settings? • Are there authentic ways to build community, even in asynchronous courses? • Will students feel they are missing out on peer interactions if an online course is asynchronous?

Recommendations provided in this chapter are grounded in student-generated data from an established online chemistry course for nonmajors, the Chemical World: The Chemistry of Bears. While this is an asynchronous course, there are consistent deadlines for assignments three times per week. As a summer course, it follows a compressed timeline of five weeks but is equivalent in instructional time and workload to a 15-week semester course. Course content includes chemical concepts (*e.g.*, conservation of mass, atomic theory, ideal gases), lab skills (*e.g.*, data collection and interpretation), and real-world applications (*e.g.*, astrochemistry, nanochemistry, and cosmetic chemistry). A course syllabus is available in the Supporting Info of a prior publication (17).

Bears are used as the guiding course theme because it is the institution's mascot, but other mascots, particularly animal ones, could readily be swapped in. The bear theme is incorporated into weekly modules such as "Feed the Bear" and "Save the Bear," with each module addressing relevant chemistry topics. Others have reported successful use of themed courses for nonmajors chemistry classes; some recently published examples of course themes include pop culture (18–21), medicinal chemistry (22, 23), human health (24) and food chemistry (25–27).

The following sections include recommendations for teaching chemistry online, informed by analyses of student response data (approved by Mercer University's Institutional Review Board), instructor observations, and pedagogically-informed approaches from an instructional designer. Readers are also encouraged to explore the other chapters in this text, as many discuss one or more of these areas. After presenting recommendations to address these interactions in online settings, this chapter concludes with suggestions for ways to apply these recommendations for fully online classes to face-to-face and hybrid settings.

Student-Instructor Interactions

There are many ways that instructors can promote positive interactions with their online students. The first day of class, or, for asynchronous courses, the opening of the course, provides an opportunity to outline clear class communication expectations that promote positive interactions for the duration of the course. Setting and maintaining clear communication strategies in terms of frequency, mode, and purpose provides students with clarity regarding instructor availability, signals to students that their instructor is actively engaged, and establishes clear expectations related to student and instructor interactions.

First Day of Class

Consider the first class meeting of the term. The first class meeting can be key to establishing class rapport and connection both to and between students, no matter what level of chemistry course. This dynamic can be challenging to replicate in an online setting. For example, what happens when the first day of class no longer means going into a physical classroom? If you are able to hold a synchronous class meeting, and if technology permits, you might be able to mimic an in-class session by creating a space where students can still see and interact with their professor and peers in real time. For asynchronous courses, this solution is not feasible; however, this does not mean instructors teaching asynchronously must lower expectations and settle for less engagement and interaction.

For example, the introductions typically made in an in-person class can translate well to a short instructor introduction video. This video provides students with a face to go with the name of their instructor and signifies that the instructor is actively engaged in the course and there to promote student success. A welcome note with an instructor photo can be used in place of an introduction video.

Giving a low stakes assignment early in the course that asks students to share introductory information can help set an early example of communication with the instructor (Figure 1). Some might wonder why preferred pronouns and pronunciations are useful if a class is asynchronous; the authors have found asking for such information helps convey to students that they want to know more about them, even if interaction will be asynchronous. The authors have successfully implemented assignments like this using a variety of modes in online courses, such as surveys and discussion boards. Discussion boards can provide an opportunity to build both student-instructor rapport and student-peer interactions and are discussed more below. If used as a discussion board assignment as illustrated in Figure 1, students can practice interacting with their peers and this can be used as a practice for future student-peer interactions. While there are many different options for an introduction assignment, an assignment such as this delivered through a discussion board can both promote student-instructor interactions and student interactions with their classmates.

> **Instructions:**
> *Let's get to know one another! Remember the overall guidelines for discussion posts in the course syllabus: after making your post, you should read the post <u>directly above yours</u> and make a thoughtful reply. You must complete your post first and then check to see who is directly above yours.*
>
> In your POST, you should address the following:
> (1) Your preferred first name, (2) where you are taking The Chemistry of Bears this summer (and why in that particular location), (3) and two additional trivia facts about yourself. Please share your preferred pronouns if you would like. Be sure to sign your post so we can all practice using preferred names. (Minimum length: 4 sentences)
>
> For your RESPONSE post, you should make a substantive response to your peer's trivia facts using their preferred first name. For example, if I were replying to a post above mine that talked about experience with travel, I might share a story of my own from a city we have both visited. (Minimum length: 4 sentences)

Figure 1. Example of a student introduction assignment via a discussion board.

Communication with Students

While the first day of class sets the tone for communication, regardless of course delivery mode, it is still important in an online class to convey clearly to students how, when, and why an instructor will communicate with them, and vice versa. Some specific considerations to make with respect to online communication are provided in Table 2.

Table 2. Guide for communication plan development for online courses.

Area to consider	*Examples*
Communication modes/platforms	*Email, learning management system, third-party platforms*
Frequency of student contact	*Daily, weekly*
Response time to student inquiries	*Within 24 h, weekend policy*
Expectations for student communication	*When, why, and how students should contact the instructor*

Table 2 provides a framework to designing a course communication plan that best fits the needs and preferences of the course and instructor. While some institutions may already have set requirements for communication with students, most instructors have significant freedom in deciding course guidelines. Although a communication plan is also important for face-to-face classes, these students likely already have more opportunities to ask questions and receive reminders because they regularly see their instructor. When a class is online, particularly when it is running without synchronous class sessions, students need clarity on what ways an instructor will communicate with them about class. Using the learning management system (LMS) announcement feature can be a good way to keep a permanent record of course information, but students should be encouraged to turn on email notifications for new announcements. If an instructor prefers to use email for class announcements, that should be conveyed to students. Additional third-party platforms might also provide engaging and timely communication strategies; however, instructors are encouraged to check with their institution regarding third-party platform policies.

Similarly, unless an institution or department has a specific required frequency of contact with online students, instructors generally can decide how often to make class announcements. For full quarter or semester classes, a weekly announcement that provides an overview of course activities and assignments might be appropriate, while a more compressed course might require communication that is more frequent. For example, since The Chemistry of Bears is on a five-week compressed timeline, one LMS announcement is posted daily. Course announcements generally include reminders about due dates, some additional real-world connections to current course topics, and class-wide feedback on a recently graded assessment.

Finally, instructors should share clear guidelines and expectations in regards to how quickly they will respond to student emails or messages as well as for what reasons students should reach out. While it might seem obvious to a seasoned instructor when and why a student should contact them, students often do not realize instructors are available for questions, concerns, and even other issues like advising and career advice. As with communication modes, some institutions have set guidelines for response time to student questions, but many instructors can set this policy themselves. It is strongly recommended to select a response time and window that will be adhered to throughout the length of the course. While KDK uses a 12 h response time, this is not a fit for every instructor; instructors should choose the response time that works best for them and their course, describe it to their students, and stick to it as much as possible. Previous work found that nonmajors online students had no preference for a specific communication plan as long as instructors (1) clearly described the plan to students and (2) adhered to the communication mode(s), frequencies, response times, and expectations throughout the course (*17*).

After deciding the how, when, and why to communicate with online students, this plan should be clearly articulated in the course syllabus, conveyed to students on the first day of class (or equivalent), and sent out in occasional reminders. This is particularly important for students who are completing labs asynchronously so that they know when to expect responses to questions and concerns about labs they are performing independently. Similarly, online students need to be reminded of when and why they should contact the instructor. In a face-to-face class, students frequently are asked "What questions do you have?" Replicating this in synchronous online courses is straightforward, as it can be asked in real time course delivery. For asynchronous courses, instructors need to be intentional about prompting students for questions in their interactions with them, including in course announcements, assignment instructions, and assessment feedback. Because students will not see the instructor at set times in an asynchronous online course, it is even more important that instructors frequently remind students of their availability for questions and concerns.

Another specific communication strategy that translates well to online courses is a midterm check-in. A midterm check-in with students that welcomes course feedback demonstrates the instructor's commitment to fostering learning and a positive course environment. A midterm check-in is administered as an anonymous but graded survey in The Chemistry of Bears; students receive their quiz points if the survey is completed. While midterm check-ins do not have to be graded, they do seem to generate more helpful feedback when they are anonymous submissions. Questions used on the midterm survey are in Figure 2.

> **Questions:**
> Q1. What is the instructor doing that is helpful for your learning?
> Q2. What can the instructor do to improve this course?
> Q3. What are you (the student) doing to be successful in the course?
> Q4. What course materials are most helpful to your learning? Most fun? Should anything be dropped in future course offerings?
> Q5. What topics would you be interested in learning about in this class in the coming weeks? (You can list topics already on the syllabus or additional topics—if possible I will incorporate some of your ideas in our course material.)
> Q6. Have you experienced any technology challenges while taking this course?
> Q7. What course components are helping you engage with other students?
> Q8. What course components and/or strategies might be used to help improve engagement with course material? How about with other students?

Figure 2. Example of a midterm check-in.

Student Interactions with Course Content

In face-to-face classes, instructors set the pace of the course and decide in what ways students will interact with material. The instructor may use a range of pedagogical moves, but ultimately, they are the revealer of course content. In contrast, in online classes, students generally have far more choice in how they move through a course, even when course materials are released one module at a time and/or at set time increments. Instructors should consider the following while designing their online course: (1) how to ensure desired progression through course material; (2) how the course workload matches that of in-person instruction; and, (3) how student learning will be assessed in the online environment. Each of these factors are important contributors to online student learning and engagement and are discussed in more depth in the following sections.

Engagement with Course Content: Course Progression

Instructors often have freedom over the content and progression of their online course. What will students learn? What will they do? How and when should students engage with material? These questions can help direct the design of a new online course.

A good organizational tool for designing a new online course is a course plan, which is a guiding document that helps an instructor tie each learning objective to specific course activities, including assessments. Starting from the main course objectives, specific learning objectives should be written that are then mapped to class content and required activities. Assigning due dates for each activity prior to starting a course enables the instructor to anticipate workload on the students and grading workload on themselves. A sample course plan is available for download as part of a previously published work (17).

The completed course plan then guides the assembly of the online course using the tools of the institution's LMS. Specific LMS companies have different terms for spaces for course materials, such as modules or pages, but the central question is the same regardless of LMS used: how does the instructor want the students to access the material? Will it all be available from the first day of the term, or will it be released at set intervals? Will there be threshold activities that require students to demonstrate mastery of a topic before receiving access to the next set of class material, or can students complete released material in any order? Providing students with clear guidance for what order they should complete course activities can help keep them on track and should promote course engagement.

Depending on previous experiences, a student may expect an online course to function like a correspondence course; that is, one where all activities can be completed any time as long as they are turned in prior to the end of the term. Other students might not realize that the course requires additional non-graded work; this is typically a bigger issue in asynchronous courses where students do not have set class meeting times with delivered course content. With students entering online courses with a range of past experiences and expectations, an overview of the course structure, including where to find activities and deadlines, helps orient students to the class and promotes a productive course start. A core strategy in increasing student engagement with course material and ensuring students submit their coursework on time is for instructors to set deadlines within the LMS for all assigned work. This will populate a checklist and/or calendar of items with all coursework due dates for students to view and follow. In a face-to-face course, assessments delivered in class such as quizzes, exams, and labs will only be missed by students if they do not attend the lab meetings. In contrast, in asynchronous online courses, students typically do not have the same frequency of prompting from the instructor or a specific class meeting time as a reminder of assessments. If due dates and times are not associated with assignments within the LMS, students may overlook work that you want them to complete and this is likely to reduce their overall engagement with course material.

Providing students with additional reminders of assignment schedules can help promote student success in online courses. The completed course plan can be a template to develop a supplemental guide for navigating students through a course. This might take the form of an orientation document, video, checklist, or other form—there are many ways to help students understand the structure of the online course. The Chemistry of Bears students have found weekly one-page guides to course activities and assessments helpful for staying on track with coursework, and the syllabus also includes a checklist of all course assessments with due dates.

Balancing Workload

Students are sometimes surprised by how much longer learning tasks can take when performed independently rather than when delivered by an instructor. For example, in a face-to-face chemistry lab session, instructors are available while students work through a lab activity. When lab is performed asynchronously, however, students will not have the opportunity to stop at each procedure step to ask for guidance, a prospect that can be intimidating to students, particularly nonmajor students. It will take a student longer to reason through the steps themselves than to have the direct, real-time guidance of an instructor or even peer. This does not mean instructors are not available for help in asynchronous online courses, but it does mean both instructor and students need to be on the same page for how long to anticipate devoting to independent lab activities. Other online learning activities require similar transparency of time expectations. For example, instructors who routinely incorporate active learning tasks into face-to-face class meetings in place of traditional lecture have found that students will often take longer to grapple with material themselves compared to passively listening to a lecture. Similarly, students will take longer working through material independently in online courses. Thus, care should be taken to make accurate estimations of times for each activity in an online course and make these time requirements transparent to students.

For in-person courses, workload estimations can be straightforward as most of the content and assessments are delivered synchronously so there are clear cues to students for how long quizzes, exams, and labs should take. Instructors likewise can observe in real time how long students take to complete these assessments and other learning activities like group work or class discussion. Synchronously-delivered online courses and hybrid courses may offer similar benefits. However, since asynchronous online courses lack set meeting times, students may struggle to understand the time investment required not only to complete assessments but also the time required to learn course material for successful assessment completion. Conveying how the work in an asynchronous course translates to the face-to-face course equivalencies of direct instruction and out-of-class time will help students better understand the time investment required for an online course.

While some institutions publish their own guidelines for calculating online course workload, many instructors are left to determine instructional equivalencies on their own. How long should one allot for a discussion board post? For reading course materials? Fortunately, there are published metrics for calculating workload equivalencies (28–30), and instructors are encouraged to use them to better reflect on course workload on students. For example, a typical Chemistry of Bears discussion board where a student makes a post, reads the peer post above theirs, and then responds to it will take students 20-25 minutes to complete, assuming a total word count of 100 words (29). Discussion boards that require students to read all peer posts and responses before posting would have a longer instructional equivalency time.

It is equally important to consider the workload on oneself as the instructor, and this is where pre-planning with a course map can really pay off. While this is often obvious when teaching in-person, it can be all too easy to forget when teaching online that even if one does not have synchronous class meetings, each assignment that comes in should serve a purpose—and will take time to grade. Finding ways to leverage the benefits of the LMS can save instructor time and sanity. For example, making use of rubrics in the LMS provides clarity to students about feedback while also making grading more efficient for the instructor. For low-stakes assignments, consider how grading schemes might promote faster feedback. For discussion posts in The Chemistry of Bears, three points are awarded for each assignment: typically, one point for the peer response and two points for the post. This allows the instructor to focus on written feedback to the posts after relatively quick assignment of points. Many LMS have the option to reuse commonly needed feedback; for example, some LMS allow the instructor to save comments when using a rubric in the free-form comment mode. An instructor might find it straightforward to copy and paste common feedback from another document into the LMS. These practices still provide helpful feedback to students but help minimize grading burden on the instructor.

Online Assessments of Student Learning

As with face-to-face classes, the best assessments are the ones that give both the instructor and students feedback on student learning. If a course plan is used in course design, the assessments should, just as for an in-person class, map to specific course objectives and goals. No work should be busy work; all activities should be purposeful. Discussion boards in particular can feel pointless to students if they are not clearly tied to course material and objectives. While one does not need to list every learning objective on every assessment—unless institutional guidelines require it —giving students some cues of assignment purpose can help students better focus their learning. Graded student work in The Chemistry of Bears includes discussion boards, reading quizzes, and lab project reports, and the assignment directions and daily announcements for each assessment remind students of the purpose for each activity. This short reminder helps highlight connections to course material as well as the skills that are being practiced. To prepare for these assessments, students work through the course reading, which includes written and video material, and complete lab experiments.

In this five-week course, deadlines fall on Mondays, Wednesdays, and Sundays, and generally student work is graded and returned prior to the next deadline. This continuous stream of formative feedback ensures that students are well-informed of their course progress, but this accelerated assessment schedule is not necessary for every online class. Communicating return times on assessments can help students understand when feedback will be received, so instructors are encouraged to share their anticipated grading times with students and clearly communicate when significant grading delays occur.

Students complete lab activities on their own in a combination of hands-on lab exercises and virtual simulations. The hands-on lab exercises for the nonmajors course are from Carolina Distance Learning (*31*) and provide students experience with standard lab equipment such as balances, test tubes, and graduated cylinders while being a minimal time burden on the instructor: the company builds the lab kits and mails them directly to the students and also provides full lab procedures and suggestions for lab report questions. Other lab assignments are completed using virtual lab simulations from PhET Interactive Simulations (*32*). While this combination works well for The Chemistry of Bears lab experiences, there are other lab approaches that have been reported in the literature, with examples including using virtual reality (*33, 34*), assembled lab kits (*35–38*), videos of in-person experiments (*39, 40*), online repositories (*41–43*), and household/food items (*27, 44, 45*). Additional examples of ways to address chemistry lab skills in an online setting, including data analysis and collaboration, have been reported by others in a special edition of the *Journal of Chemical Education* (*46*).

The laboratory learning outcomes for The Chemistry of Bears are provided in Figure 3. These learning outcomes are all assessable with written lab reports, which is essential due to the asynchronous delivery of the course. Without being in the physical lab or even on video conferencing with a student, the instructor cannot monitor lab behaviors in real time. Instead, students are asked to take pictures of key moments in the lab activity; this also could be requested as short videos.

Students will be able to:
1. Generate a hypothesis to explain natural phenomena;
2. Collect and organize experimental data in a format appropriate to a scientific field;
3. Analyze data through the use of quantitative and/or qualitative scientific reasoning;
4. Interpret a hypothesis in light of experimental evidence;
5. Accurately communicate scientific knowledge, observations, analyses, and/or conclusions.

Figure 3. Sample nonmajors chemistry lab learning outcomes.

The outcomes in Figure 3 are assessed via post-lab reports rich in reflection prompts. For example, students are asked to develop a hypothesis statement prior to starting a lab and then, after completing all data collection, are asked to explain if their hypothesis was correct or not, using their data to support their answer. In early labs, students are provided with blank tables and charts to fill in their data and observations, but later labs require them to decide how best to present their results. Since the instructor is not observing the students in real time, an additional question on each lab report asks students to share how the lab activity went and how long it took. This provides the instructor with additional opportunities to connect with students via feedback specific to how well students perceived a lab activity to go. For example, early in the course students often express being nervous about messing up lab procedures; this gives the instructor the opportunity to share individual and class-wide encouragement. Moments like these are typically unplanned in synchronous and face-to-face courses as they organically happen, so instructors of asynchronous online courses should be aware that more intentionality might be needed to help students navigate lab activity concerns.

Academic Integrity

Academic integrity can be a perennial classroom concern, and opinions vary widely on best practices for monitoring (or not monitoring) students during online exams (47–50). One approach to avoid the perceived need for online proctoring is to allow notes and other class resources to be used in all assessments, including online quizzes and exams. Students in The Chemistry of Bears take twice-weekly reading quizzes that are untimed and open-note but with set deadlines. One requirement that has helped focus students' attention to the class materials instead of broader internet searches is to state that answers "should be based on class materials only." This cue helps remind students of the scope of the assessment as well as allows the instructor to focus their grading on answers relevant to material covered in the course. The other assessments—discussion boards and lab activities—use the same requirement of basing answers on class materials only. This has worked well to direct students to course materials while also allowing for assessment of their learning. Other approaches that help students focus their time on class materials instead of internet searches include embedding short quizzes between videos and readings and referencing source material directly in quiz questions.

One issue that can be of concern is how to ensure that students are completing their own lab work in an asynchronous course. This is addressed in several ways in The Chemistry of Bears. It is made clear to students after registration, on the syllabus, and in several announcements that all students are required to purchase their own set of lab materials and work independently on their assignments. On the actual lab reports, students often are asked to take selfies with various lab equipment; this is particularly useful for checking that they are using their required safety equipment. Pictures of results are taken on a "lab placemat," which is a piece of paper with a heading specific to the lab and the students' signature. This requirement is not time intensive and is an easy reminder for students of the honor code that is in place for the course.

Elsewhere in this text there are additional discussions of academic integrity issues in online courses. Readers are encouraged to explore those chapters for additional approaches for their own courses.

Student-Peer Interactions

One common concern with online instruction is how to replicate the in-person camaraderie and level of engagement in what might seem like a more static setting. Even fully asynchronous courses can build class rapport and promote interactions between students. Common approaches include effective use of discussion boards and incorporation of group work. Implementation of these approaches in The Chemistry of Bears are discussed below as examples.

Utility of Discussion Boards

Students in The Chemistry of Bears complete at least ten discussion board prompts over the five-week compressed course. While a common approach to discussion boards is to require each student to write one initial post followed by two replies to posts of fellow classmates, this can be frustrating to proactive students who are forced to wait on their peers to complete work in order to generate enough posts for replies. Additionally, students often report wasting time scrolling through posts to find potential prompts for replies, or, perhaps worse, only giving cursory attention to the assignment. To avoid these discussion board pitfalls, The Chemistry of Bears students are asked to post once and then reply to the post directly above theirs. An example of these discussion board instructions is provided in Figure 1. This posting policy attempts to mimic an in-person class discussion; most typically in a face-to-face discussion of a text or concept, students respond to a point made by the person who most recently spoke. An additional benefit to this approach is that students who are proactive about completing their discussion board assignment do not have to wait for others to post: if they are the first to post, they are done with the assignment.

For asynchronous courses, discussion boards may be the main way that students will interact with their peers. The Chemistry of Bears students frequently note the benefit of discussion boards for building connections with peers. A representative student comment captures this sentiment: "*Being able to respond to other students in a fun way makes it feel as if we are actually in a class together and not just [in] another online class where you do not know your peers.*"

Group Work

Others have reported effective approaches to group work in online settings. Giving students opportunities to work together enhances learning while also developing written and oral scientific communication skills (*51*). For example, video conferencing has been used to match students with learners at other institutions to work together on organic chemistry problems (*52, 53*). Online group work also can be effective in asynchronous formats, as demonstrated by Winschel *et al.* in a publication on online spectroscopy problems (*54*). The 2020 pivot to remote instruction necessitated rapid, creative solutions to preserve face-to-face group work assignments, and some are already reporting on their approaches, which include group exams (*55*), synchronous group learning (*56–59*), and asynchronous group activities (*60, 61*).

These examples are just a small sampling of possibilities that could aid online instructors in the design of group work activities. As with any pedagogical approach, instructors should consider if it is a fit for their class. If the group work will be synchronous, such as requiring students to work collaboratively in web conferencing tools, is it falling during a scheduled synchronous class meeting time? If not, how can the instructor ensure that all students will be available? Is the learning occurring in the group work only possible through synchronous collaboration, or are there ways to achieve the same learning outcomes through asynchronous platforms such as discussion boards, email, or collaborative document spaces such as Google Drive, Discord, or Slack? The practicalities of implementation become far more significant in an online setting than in a face-to-face class where students and instructors can work through technology and content issues together.

If group work is designed to be part of an online class, it is recommended that instructors provide a mechanism for students to provide private feedback to their instructor about their group work contributions, both those of their groupmates as well as their own. This can be done via informal surveys, more formal written reports, or oral interviews. Whatever the format, asking students to reflect on group work contributions can help instructors better understand group dynamics and products and supplement their own observations—a particularly important aspect if the group work is performed asynchronously.

What Do Students Want in an Online Course? Ask Them.

What do students want in an online course? The answer is, often, "it depends." One method to gain insight into students' wants and expectations for online coursework is to survey students during and after the course with specific questions about course design, content, engagement strategies, instructor availability, and other topics of interest to the instructor. Survey responses provide real-time feedback on areas of success and opportunities for improvement. Here we provide an example of how such feedback could be used in an online course by discussing survey responses from students enrolled in the 2020 iteration of The Chemistry of Bears.

Halfway through the course, students were given an anonymous, graded survey in the LMS with the questions in Figure 2. To encourage survey completion, the survey counted as a quiz grade. Here we focus on analysis of three of the survey questions:

Q4. What course materials are most helpful to your learning? Most fun? Should anything be dropped in future course offerings?

Q7. What course components are helping you engage with other students?

Q8. What course components and/or strategies might be used to help improve engagement with course material? How about with other students?

Discussion boards were cited in 98% of survey responses as the most helpful course component for engaging with their peers. While it is not unusual for students to informally express concerns about the usefulness of discussion boards prior to the start of class, it is clear that most found them beneficial with respect to promoting engagement. Interestingly, when asked what course components were most helpful and what could be dropped, 23% of class responses described the discussion boards as the most helpful course component for learning, while 7% of respondents suggested dropping them.

Students also had suggestions for new course approaches that would promote engagement with course peers and material. Some responses cited group work as a desired new course approach for benefiting their learning. As discussed earlier, group work is not already part of the course. Interestingly, most of these comments specified adapting just one lab project to be a group experience, with most citing the discussion of lab results as being the most potentially beneficial to their learning. This suggestion was not incorporated in the latter half of the course, but it is being explored for future course iterations.

Another peer engagement suggestion came from 14% of respondents: incorporate periodic live video chats as a way that students could better engage with one another, the instructor, and course material. The Chemistry of Bears is intentionally asynchronous as enrolled students typically have a range of summer obligations, and a common meeting time has not been found for student interaction sessions. That said, some courses might benefit from occasional live meetings, and these sessions could be recorded and viewed later. Similarly, instructors might find optional online office hours or review sessions to be a good fit for their online or hybrid courses.

To summarize, the midterm survey in The Chemistry of Bears generated the following key results from 43 responses:

- Discussion boards are an effective tool to promote engagement and learning.
- Group work could benefit learning, particularly in lab activities.
- Instructors might consider incorporating synchronous moments into asynchronous courses.

It may be that some of these results are transferable to other chemistry courses and even other institutions, but the main goal of the above discussion was to serve as an illustration of potential insights gained from surveying students during an in-progress course. Online instructors are encouraged to adapt the midterm check-in questions as appropriate to their own courses, with the intention of generating student feedback while the course is in progress. Analysis of responses may also help instructors identify places where additional feedback should be provided to the class. For example, responses about technology issues might help an instructor identify solutions for students that could be shared via class announcements.

Additional Considerations for Online Laboratory Courses

Teaching chemistry labs via online settings presents additional challenges and concerns, particularly when the course is taught asynchronously. When labs meet face-to-face or even synchronously online, the instructor can readily see who is in attendance. Similarly, these modalities give easy access to the instructor for help in real time. This enables the instructor to easily correct lab technique as the lab is being performed, answer questions about the procedure, and even give full-class feedback as necessary. For asynchronously-delivered lab courses, instructors should anticipate, as best they can, potential problem areas before students start each lab. For example, in order to identify potential areas of confusion, the Chemistry of Bears labs were tested ahead of the course delivery by volunteers that had minimal chemistry experience. Still, each year the course is taught new questions about labs are brought up by students. In efforts to address these questions and potential problems proactively, we suggest keeping a running list of repeated question areas, lab feedback, potential problems and pitfalls, etc. This feedback can be used for the creation of a robust "Tips & Tricks" page in the LMS with clarification and hints on lab topics ranging from proper equipment use to examples of appropriate observations.

Another area that can be challenging in asynchronous chemistry courses is lab attendance, since students complete lab activities on their own and not during scheduled lab times. Thus, the instructor may only know that a student has "shown up" for lab once they submit their lab assessment(s) for grading. There are myriad reasons for why students do not complete lab work for in-person classes, but online labs present a barrier typically not present for in-person classes: the cost of lab supplies. While in-person laboratories have all resources needed for students to complete lab—except perhaps for lab notebooks and safety equipment—if online labs are hands-on, students must obtain their own lab materials. For online lab courses that require students to purchase household items to complete, instructors are encouraged to provide a list of all items and their anticipated cost at the start of the semester so that students are not surprised by cost. For The Chemistry of Bears, which requires students to purchase lab equipment and supplies from a company, students are advised of the cost at the time of registration so that preparations can be made ahead of the start of class.

Application to In-Person and Hybrid Classes

Recommendations made in the previous section can also be applied to hybrid and in-person courses. Finding ways for students to work together creates opportunities for students to build community while constructing new knowledge. While the setting and mode for building community might change—from laboratory interactions, engaging class discussions, or even asynchronous online discussion boards—the benefits persist.

After the abrupt shift from in-person instruction to remote instruction in Spring 2020 and a small in-person summer pilot, Mercer University returned to in-person instruction for the Fall 2020 semester, with necessary masking, hygiene, and social distancing requirements in place due to COVID-19. While the return to in-person instruction brought a partial return to "normal," the COVID-19 guidelines and restrictions created a need to modify and adapt once again. For example, due to social distancing, group work, including partnered lab work, was not feasible. However, despite these restrictions, in-class engagement was still possible, and approaches from online classes can help enable students to interact with course content, their peers, and even the instructor.

For example, co-author LES successfully used Google Forms to enable real-time feedback to students during in-person sociology classes. Students are shown a QR code on the screen that links to a pre-constructed Google Form with prompts related to the reading. As students submit answers, LES can quickly assess overall class understanding and use specific student responses to move class discussion forward and/or address class misconceptions. A slight modification to this approach was implemented in synchronous Zoom sessions in hybrid courses where students were put into breakout rooms and tasked with working together to respond to provided discussion prompts. After completing their work, the group provided a shared response via Google Forms. Submissions were compiled and shared with the full class for further discussion and demonstration of the different approaches and responses across each group.

Google Forms for group work reporting out also translates well to hybrid courses in chemistry. For example, KDK has used this approach in a sophomore-level quantitative analysis course that functioned as a hybrid course with each week consisting of a 75-minute in-person lecture, a 75-minute Zoom session for group work, and a three-hour in-person laboratory period. In the Zoom sessions, in efforts to better mimic in-person group work that typically has strong participation from all group members, students were asked to select specific roles for each group discussion session (see Figure 4). The Google Form used by each group included places to record each group member's role, so the instructor could easily log attendance and monitor how well groups rotated through each role. The roles also helped to keep groups on-task, even when the instructor was not present in the breakout room. The group roles used in Figure 4 are equally appropriate for in-class group work, even though in-person modalities give instructors a better sense of class participation than when students are in separated online video conference spaces.

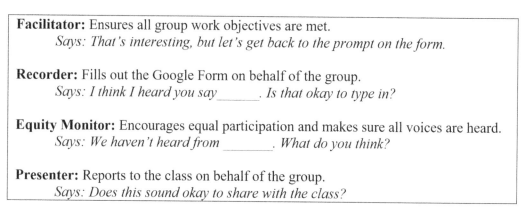

Figure 4. Group role descriptions and examples of interactions.

Another approach used in online classes that can be beneficial for in-person and/or hybrid classes are discussion boards, as they can be a productive space for engaging students in discussion when an in-person class session does not permit this due to class size, meeting length, and/or course material-induced restraints. Of course, just as with online courses, it is important that discussion boards are thoughtfully designed to be integral to achieving course outcomes and not seen as busy work that is unnecessary to the class. Online discussion boards also have utility as non-graded spaces for students to ask questions, share lab results, or address other class-related topics.

Conclusions and Future Work

The recommendations discussed throughout this chapter provide ways to address three types of student interactions when teaching online: interactions with their instructor, with course content, and with fellow online classmates. While the suggestions shared herein are from observations and data from The Chemistry of Bears and its corresponding population, they likely have broader applicability for other online science courses. Furthermore, given the ongoing expectations in regards to multiple modes of course delivery, we have included examples of how online methods can be integrated or adapted for in-person and hybrid courses and vice versa.

It is important to note that these recommendations come after four years of teaching a fully online chemistry course for nonmajors that was designed with a full year of preparation time prior to course delivery and with significant guidance from an experienced instructional designer. Just as with in-person offerings, online courses will evolve and mature the more iterations they are taught. Online courses with lab components bring added challenges, and even with large lead times for preparation, unexpected issues can crop up when working with students remotely, as discussed earlier in the chapter. Instructors new to online teaching can hopefully feel encouraged by our experiences with this course in that it does indeed get easier and smoother with each iteration.

There are opportunities to learn more about student expectations for online learning. With the rise of use and comfort with web conferencing, do students now expect that online classes should include synchronous components? Should online instructors anticipate requests for video meetings with their online students? How will this generation of students be affected by their remote learning experiences? These questions would benefit from additional study as they will likely provide insight to student learning in online settings and may be helpful to consider in designing and revising of future online chemistry courses.

Acknowledgments

Thank you to Dr. Erin Whitteck (University of Missouri—St. Louis) for helpful discussions. The authors also thank Dr. Gretchen Potts (University of Tennessee at Chattanooga) for sharing online teaching materials and approaches. Madison Reyome provided help with proofreading. Finally, the authors thank the students of CHM 110 for their survey participation, and the anonymous reviewers for their thoughtful feedback.

References

1. Holme, T. A. Chemistry Education in Times of Disruption and the Times That Lie Beyond. *J. Chem. Educ.* **2020**, 97 (5), 1219–1220. https://doi.org/10.1021/acs.jchemed.0c00377.
2. Holme, T. A. Will 2020 Be an Inflection Point in the Trajectory of Chemistry Teaching and Learning? *J. Chem. Educ.* **2020**, 97 (12), 4215–4216. https://doi.org/10.1021/acs.jchemed.0c01396.
3. Major, C. H. *Teaching Online: A Guide to Theory, Research, and Practice*; Johns Hopkins University Press: Baltimore, MD, 2015.
4. Flynn, A.; Kerr, J. *Remote Teaching: A Practical Guide with Tools, Tips, and Techniques*; Pressbooks: 2020. https://ecampusontario.pressbooks.pub/remotecourse.
5. National Research Council. *Discipline-Based Education Research: Understanding and Improving Learning in Undergraduate Science and Engineering*; National Academies Press: Washington, D.C., 2012.

6. Hilosky, A.; Sutman, F.; Schmuckler, J. Is Laboratory Based Instruction in Beginning College-Level Chemistry Worth the Effort and Expense? *J. Chem. Educ.* **1998**, *75* (1), 100–104. https://doi.org/10.1021/ed075p100.
7. Hawkes, S. J. Chemistry Is Not a Laboratory Science. *J. Chem. Educ.* **2004**, *81* (9), 1257. https://doi.org/10.1021/ed081p1257.
8. Hofstein, A.; Lunetta, V. N. The Laboratory in Science Education: Foundations for the Twenty-First Century. *Science Education* **2004**, *88* (1), 28–54. https://doi.org/10.1002/sce.10106.
9. Reid, N.; Shah, I. The Role of Laboratory Work in University Chemistry. *Chem. Educ. Res. Pract.* **2007**, *8* (2), 172–185. https://doi.org/10.1039/B5RP90026C.
10. Elliott, M. J.; Stewart, K. K.; Lagowski, J. J. The Role of the Laboratory in Chemistry Instruction. *J. Chem. Educ.* **2008**, *85* (1), 145–149. https://doi.org/10.1021/ed085p145.
11. Bretz, S. L. Evidence for the Importance of Laboratory Courses. *J. Chem. Educ.* **2019**, *96* (2), 193–195. https://doi.org/10.1021/acs.jchemed.8b00874.
12. Seery, M. K. Establishing the Laboratory as the Place to Learn How to Do Chemistry. *J. Chem. Educ.* **2020**, *97* (6), 1511–1514. https://doi.org/10.1021/acs.jchemed.9b00764.
13. Garrison, D. R.; Anderson, T.; Archer, W. Critical Inquiry in a Text-Based Environment: Computer Conferencing in Higher Education. *The Internet and Higher Education* **1999**, *2* (2), 87–105. https://doi.org/10.1016/S1096-7516(00)00016-6.
14. Garrison, D. R. Online Community of Inquiry Review: Social, Cognitive, and Teaching Presence Issues. *Journal of Asynchronous Learning Networks* **2007**, *11* (1), 61–72.
15. *The Community of Inquiry*. https://coi.athabascau.ca/ (accessed 2021-06-30).
16. *Applying the Community of Inquiry Framework*. https://cte.virginia.edu/resources/applying-community-inquiry-framework (accessed 2021-06-30).
17. Simon, L. E.; Genova, L. E.; Kloepper, M. L. O.; Kloepper, K. D. Learning Postdisruption: Lessons from Students in a Fully Online Nonmajors Laboratory Course. *J. Chem. Educ.* **2020**, *97* (9), 2430–2438. https://doi.org/10.1021/acs.jchemed.0c00778.
18. Hickey, S. P. *Game of Thrones*, *Breaking Bad*, Nicolas Cage, *Harry Potter*, *Pulp Fiction*, and More: The Key Ingredients in Teaching Biochemistry to Nonscience Majors. In *Videos in Chemistry Education: Applications of Interactive Tools*; Parr, J., Ed.; ACS Symposium Series, Vol. 1325; American Chemical Society, 2019; pp 1–19. https://doi.org/10.1021/bk-2019-1325.ch001.
19. Tallman, K. A. Introducing Students to Fundamental Chemistry Concepts and Basic Research through a Chemistry of Fashion Course for Nonscience Majors. *J. Chem. Educ.* **2019**, *96* (9), 1906–1913. https://doi.org/10.1021/acs.jchemed.8b00826.
20. Hickey, S. P. Can You Teach Subatomic Particles with *WKRP in Cincinnati* and Climate Change with *Last Week Tonight with John Oliver*: Conveying Chemistry to Nonscience Majors Using Videos. In *Communication in Chemistry*; Crawford, G. L., Kloepper, K. D., Meyers, J. J., Singiser, R. H., Eds.; ACS Symposium Series, Vol. 1327; American Chemical Society, 2019; pp 163–185. https://doi.org/10.1021/bk-2019-1327.ch012.
21. Avila-Bront, L. G. An Experiential Learning Chemistry Course for Nonmajors Taught through the Lens of Science Fiction. *J. Chem. Educ.* **2020**, *97* (10), 3588–3594. https://doi.org/10.1021/acs.jchemed.0c00264.

22. Neuman, A. W.; Harmon, B. B. Plants in Medicine: An Integrated Lab–Lecture Project for Nonscience Majors. *J. Chem. Educ.* **2019**, *96* (1), 60–65. https://doi.org/10.1021/acs.jchemed.8b00583.
23. Wenzel, A. G.; Casper, S.; Galvin, C. J.; Beck, G. E. Science and Business of Medicinal Chemistry: A "Bench-to-Bedside" Course for Nonmajors. *J. Chem. Educ.* **2020**, *97* (2), 414–420. https://doi.org/10.1021/acs.jchemed.9b00691.
24. Armstrong, D.; Poë, J. C. The Science of Human Health—A Context-Based Chemistry Course for Non-Science Majors Incorporating Systems Thinking. *J. Chem. Educ.* **2020**, *97* (11), 3957–3965. https://doi.org/10.1021/acs.jchemed.0c00887.
25. Dabrowski, J. A.; Manson McManamy, M. E. Design of Culinary Transformations: A Chemistry Course for Nonscience Majors. *J. Chem. Educ.* **2020**, *97* (5), 1283–1288. https://doi.org/10.1021/acs.jchemed.9b00964.
26. Perets, E. A.; Chabeda, D.; Gong, A. Z.; Huang, X.; Fung, T. S.; Ng, K. Y.; Bathgate, M.; Yan, E. C. Y. Impact of the Emergency Transition to Remote Teaching on Student Engagement in a Non-STEM Undergraduate Chemistry Course in the Time of COVID-19. *J. Chem. Educ.* **2020**, *97* (9), 2439–2447. https://doi.org/10.1021/acs.jchemed.0c00879.
27. Nguyen, J. G.; Keuseman, K. J. Chemistry in the Kitchen Laboratories at Home. *J. Chem. Educ.* **2020**, *97* (9), 3042–3047. https://doi.org/10.1021/acs.jchemed.0c00626.
28. Adler, K. M. Determining Carnegie Units: Student Engagement in Online Courses Without a Residential Equivalent. *Online Journal of Distance Learning Administration* **2020**, *23* (1)
29. *Workload Estimator 2.0*. https://cat.wfu.edu/resources/tools/estimator2/ (accessed 2021-06-30).
30. *Course Workload Estimator*. https://cte.rice.edu/workload (accessed 2020-06-30).
31. *Carolina Distance Learning*. https://www.carolinadistancelearning.com/ (accessed 2021-06-30).
32. *PhET Interactive Simulations*. https://phet.colorado.edu/ (accessed 2021-06-30).
33. Dunnagan, C. L.; Dannenberg, D. A.; Cuales, M. P.; Earnest, A. D.; Gurnsey, R. M.; Gallardo-Williams, M. T. Production and Evaluation of a Realistic Immersive Virtual Reality Organic Chemistry Laboratory Experience: Infrared Spectroscopy. *J. Chem. Educ.* **2020**, *97* (1), 258–262. https://doi.org/10.1021/acs.jchemed.9b00705.
34. Dunnagan, C. L.; Gallardo-Williams, M. T. Overcoming Physical Separation During COVID-19 Using Virtual Reality in Organic Chemistry Laboratories. *J. Chem. Educ.* **2020**, *97* (9), 3060–3063. https://doi.org/10.1021/acs.jchemed.0c00548.
35. Miles, D. T.; Wells, W. G. Lab-in-a-Box: A Guide for Remote Laboratory Instruction in an Instrumental Analysis Course. *J. Chem. Educ.* **2020**, *97* (9), 2971–2975. https://doi.org/10.1021/acs.jchemed.0c00709.
36. Mirowsky, J. E. Converting an Environmental Sampling Methods Lecture/Laboratory Course into an Inquiry-Based Laboratory Experience during the Transition to Distance Learning. *J. Chem. Educ.* **2020**, *97* (9), 2992–2995. https://doi.org/10.1021/acs.jchemed.0c00591.
37. Destino, J. F.; Gross, E. M.; Niemeyer, E. D.; Petrovic, S. C. Hands-on Experiences for Remotely Taught Analytical Chemistry Laboratories. *Anal Bioanal Chem* **2021**, *413* (5), 1237–1244. https://doi.org/10.1007/s00216-020-03142-1.

38. Kelley, E. W. Sample Plan for Easy, Inexpensive, Safe, and Relevant Hands-On, At-Home Wet Organic Chemistry Laboratory Activities. *J. Chem. Educ.* **2021**, *98* (5), 1622–1635. https://doi.org/10.1021/acs.jchemed.0c01172.

39. Howitz, W. J.; Thane, T. A.; Frey, T. L.; Wang, X. S.; Gonzales, J. C.; Tretbar, C. A.; Seith, D. D.; Saluga, S. J.; Lam, S.; Nguyen, M. M.; Tieu, P.; Link, R. D.; Edwards, K. D. Online in No Time: Design and Implementation of a Remote Learning First Quarter General Chemistry Laboratory and Second Quarter Organic Chemistry Laboratory. *J. Chem. Educ.* **2020**, *97* (9), 2624–2634. https://doi.org/10.1021/acs.jchemed.0c00895.

40. Woelk, K.; Whitefield, P. D. As Close as It Might Get to the Real Lab Experience—Live-Streamed Laboratory Activities. *J. Chem. Educ.* **2020**, *97* (9), 2996–3001. https://doi.org/10.1021/acs.jchemed.0c00695.

41. Benatan, E.; Dene, J.; Stewart, J. L.; Eppley, H. J.; Watson, L. A.; Geselbracht, M. J.; Williams, B. S.; Reisner, B. A.; Jamieson, E. R.; Johnson, A. R. JCE VIPEr: An Inorganic Teaching and Learning Community. *J. Chem. Educ.* **2009**, *86* (6), 766–767. https://doi.org/10.1021/ed086p766.

42. Kelly, R. S.; Larive, C. K. The Analytical Sciences Digital Library: Your Online Resource for Teaching Instrumentation. *J. Chem. Educ.* **2011**, *88* (4), 375–377. https://doi.org/10.1021/ed101108t.

43. Nataro, C.; Johnson, A. R. A Community Springs to Action to Enable Virtual Laboratory Instruction. *J. Chem. Educ.* **2020**, *97* (9), 3033–3037. https://doi.org/10.1021/acs.jchemed.0c00526.

44. Selco, J. I. Using Hands-On Chemistry Experiments While Teaching Online. *J. Chem. Educ.* **2020**, *97* (9), 2617–2623. https://doi.org/10.1021/acs.jchemed.0c00424.

45. Schultz, M.; Callahan, D. L.; Miltiadous, A. Development and Use of Kitchen Chemistry Home Practical Activities during Unanticipated Campus Closures. *J. Chem. Educ.* **2020**, *97* (9), 2678–2684. https://doi.org/10.1021/acs.jchemed.0c00620.

46. *Journal of Chemical Education* **2020**, *97* (9) https://pubs.acs.org/toc/jceda8/97/9 (accessed 2021-06-30).

47. Clark, T. M.; Callam, C. S.; Paul, N. M.; Stoltzfus, M. W.; Turner, D. Testing in the Time of COVID-19: A Sudden Transition to Unproctored Online Exams. *J. Chem. Educ.* **2020**, *97* (9), 3413–3417. https://doi.org/10.1021/acs.jchemed.0c00546.

48. Nguyen, J. G.; Keuseman, K. J.; Humston, J. J. Minimize Online Cheating for Online Assessments During COVID-19 Pandemic. *J. Chem. Educ.* **2020**, *97* (9), 3429–3435. https://doi.org/10.1021/acs.jchemed.0c00790.

49. Raje, S.; Stitzel, S. Strategies for Effective Assessments While Ensuring Academic Integrity in General Chemistry Courses during COVID-19. *J. Chem. Educ.* **2020**, *97* (9), 3436–3440. https://doi.org/10.1021/acs.jchemed.0c00797.

50. Lancaster, T.; Cotarlan, C. Contract Cheating by STEM Students through a File Sharing Website: A Covid-19 Pandemic Perspective. *Int. J. Educ. Integr.* **2021**, *17* (1), 1–16. https://doi.org/10.1007/s40979-021-00070-0.

51. Crawford, G. L.; Kloepper, K. D.; Meyers, J. J.; Singiser, R. H. Communicating Chemistry: An Introduction. In *Communication in Chemistry*; Crawford, G. L., Kloepper, K. D., Meyers, J. J.,

Singiser, R. H., Eds.; ACS Symposium Series, Vol. 1327; American Chemical Society: 2019; pp 1–15. https://doi.org/10.1021/bk-2019-1327.ch001.

52. Skagen, D.; McCollum, B.; Morsch, L.; Shokoples, B. Developing Communication Confidence and Professional Identity in Chemistry through International Online Collaborative Learning. *Chem. Educ. Res. Pract.* **2018**, *19* (2), 567–582. https://doi.org/10.1039/C7RP00220C.

53. Wentzel, M. T.; Ripley, I.; McCollum, B. M.; Morsch, L. A. Practicing Multimodal Chemistry Communication through Online Collaborative Learning. In *Communication in Chemistry*; Crawford, G. L., Kloepper, K. D., Meyers, J. J., Singiser, R. H., Eds.; ACS Symposium Series, Vol. 1327; American Chemical Society, 2019; pp 57–74. https://doi.org/10.1021/bk-2019-1327.ch005.

54. Winschel, G. A.; Everett, R. K.; Coppola, B. P.; Shultz, G. V.; Lonn, S. Using Jigsaw-Style Spectroscopy Problem-Solving To Elucidate Molecular Structure through Online Cooperative Learning. *J. Chem. Educ.* **2015**, *92* (7), 1188–1193. https://doi.org/10.1021/acs.jchemed.5b00114.

55. Goodman, A. L. Can Group Oral Exams and Team Assignments Help Create a Supportive Student Community in a Biochemistry Course for Nonmajors? *J. Chem. Educ.* **2020**, *97* (9), 3441–3445. https://doi.org/10.1021/acs.jchemed.0c00815.

56. Gemmel, P. M.; Goetz, M. K.; James, N. M.; Jesse, K. A.; Ratliff, B. J. Collaborative Learning in Chemistry: Impact of COVID-19. *J. Chem. Educ.* **2020**, *97* (9), 2899–2904. https://doi.org/10.1021/acs.jchemed.0c00713.

57. Reynders, G.; Ruder, S. M. Moving a Large-Lecture Organic POGIL Classroom to an Online Setting. *J. Chem. Educ.* **2020**, *97* (9), 3182–3187. https://doi.org/10.1021/acs.jchemed.0c00615.

58. Nickerson, L. A.; Shea, K. M. First-Semester Organic Chemistry during COVID-19: Prioritizing Group Work, Flexibility, and Student Engagement. *J. Chem. Educ.* **2020**, *97* (9), 3201–3205. https://doi.org/10.1021/acs.jchemed.0c00674.

59. Venton, B. J.; Pompano, R. R. Strategies for Enhancing Remote Student Engagement through Active Learning. *Anal Bioanal Chem* **2021**, *413* (6), 1507–1512. https://doi.org/10.1007/s00216-021-03159-0.

60. Flener-Lovitt, C.; Bailey, K.; Han, R. Using Structured Teams to Develop Social Presence in Asynchronous Chemistry Courses. *J. Chem. Educ.* **2020**, *97* (9), 2519–2525. https://doi.org/10.1021/acs.jchemed.0c00765.

61. Van Heuvelen, K. M.; Daub, G. W.; Ryswyk, H. V. Emergency Remote Instruction during the COVID-19 Pandemic Reshapes Collaborative Learning in General Chemistry. *J. Chem. Educ.* **2020**, *97* (9), 2884–2888. https://doi.org/10.1021/acs.jchemed.0c00691.

Chapter 2

Transitioning from High-Stakes to Low-Stakes Assessment for Online Courses

Matthew D. Casselman[*]

Department of Chemistry, University of California – Riverside, Riverside, California 92521, United States

[*]Email: matthew.casselman@ucr.edu

Traditional in-person lecture courses are typically assessed using high-stakes summative assessments (e.g., midterms and finals). Several existing tools exist for adapting these exams to an online environment, but they are not without drawbacks. Alternatively, reevaluating and revising the role of assessment in remote learning allows instructors to re-imagine what assessment looks like in their classroom. In this chapter, practices for transitioning from high-stakes and summative assessment to lower-stakes and formative assessments will be described in the context of a lower-division organic chemistry classroom. It will detail how students' needs are met by using frequent, lower-stakes, formative assessments in the form of online quizzes, online homework, in-class polling, and collaborative online learning activities. Strategies for designing effective summative assessments online will be discussed as well.

Introduction and Context

With the COVID-19 pandemic causing significant disruptions to higher education's standard operating procedure, a rapid transition to online instruction occurred during the spring term of 2020 (*1*). At my institution, this happened as we transitioned from the winter to spring quarter; it was uncertain how long-lasting the remote instruction requirement was going to last. As I prepared for that quarter, I considered many factors how to redesign my typical organic chemistry lecture course online:

How should lectures be structured? How could lectures be made accessible to students and remain engaging? How can we ensure students stay current with the material and prevent students from entering the cycle of cramming for exams? How will assessment differ from previous terms? How could assessment change to accommodate students in a variety of individual situations while remaining fair?

© 2021 American Chemical Society

With these questions in mind, I set out to transition to a remote version of my conventional in-person lecture course. At the same time, instructors around the world were also developing new ways to teach online effectively; these efforts were highlighted in a special issue of the *Journal of Chemistry Education* (*2*). Managing lectures themselves was a straightforward task; they could either be pre-recorded or synchronous. Synchronous lectures were employed and supported with web-based clicker polling to keep students engaged throughout the quarter. This proved to be a straightforward transition as my lecture format need not change significantly, and I had previously utilized web-based clicker polling in my classes. However, a more daunting question was how to assess students online. Should I simply adapt existing methods and strategies or redesign how assessment occurs and re-envision its role in my classroom?

To provide context for these changes, organic chemistry (CHEM 8 A/B/C series) at the University of California, Riverside (UCR) is a three-quarter (10 weeks each) sequence of courses covering the broad spectrum of typical topics included in introductory organic chemistry courses (Table 1). The laboratory component is a separate course taught by a different instructor. Courses are taught both "on-sequence" and "off-sequence." Typically, I teach "on-sequence" CHEM 8A in the fall, but "off-sequence" courses winter, spring, and summer quarters. Upon transition to remote instruction in the spring quarter, I was scheduled to teach CHEM 8B, the second course in the series; then followed with 5-week version of CHEM 8C in the summer term and CHEM 8A in the fall 2020 quarter. During the campus-wide transition to remote instruction, there were no plans to reduce the course coverage to accommodate online delivery. Organic chemistry classes at UCR are held in large lecture halls. Courses in the winter quarter before remote instruction were approximately 300-330 students in a classroom with a capacity of 330 students; class sizes online were expected to be similar in size. My in-person approach employs a modular flipped modality where approximately one out of three class periods are conducted as collaborative problem-solving sessions with lecture tutorial and clicker use. In spring and summer 2020, remote instruction featured synchronous lectures only with no flipped activities, however in fall 2020 and beyond, flipped activities were returned into typical class rotation.

Modifying Assessment for Remote Instruction

Assessment in large-enrollment lecture courses is historically dominated by high-stakes summative assessments (*i.e.,* midterm and final exams.) In the online environment, these assessments are harder to administer, and they potentially suffer from lower quality due to the reliance upon multiple-choice questions (*3*) and are plagued with concerns over academic integrity (*4*). Other instructors have found that a redesign of summative assessment in their courses has been beneficial (*5*). On the other hand, lower-stakes formative assessments are associated with greater student learning as feedback on learning progress can allow students to make adjustments in their study practices (*6*).

Because of the unique challenges of assessment in an online environment, I set out to transition students from high-stakes summative assessment to lower-stakes formative assessment (course structure summarized in Table 2). Accomplishing the transition from summative to formative assessment in remote instruction meant leveraging technology in a way that students could both submit their work for grade and receive rapid feedback on their work (*7*). The use of frequent lower-stakes assessment has been shown to be beneficial by increasing students' intrinsic motivation (*8*). The transition from summative to formative assessment may be helpful as students have been observed to learn more when engaged regularly in formative assessment (*9*). Students view formative assessment in the form of homework as beneficial to their learning for various reasons, including encouraging more frequent studying and learning from their mistakes (10). From a logistical perspective, replacing midterm exams with regular homework assignments helped free up class periods for lecturing and dealing with the slower pace of remote instruction.

Students still have regular in-class formative assessment and feedback opportunities. In the past, this included the use of web-based clickers (e.g., PollEverywhere (11)) during daily class lectures. In-class polling has been popularized in recent years and can provide feedback for students on the material they have just learned (12). In addition to in-class polling as formative assessment, collaborative learning lecture tutorials (scaffolded worksheet exercises, such as those used in POGIL (13)) are utilized to encourage discussion and engagement during class while providing valuable feedback on students' problem-solving and sense-making (14).

Table 1. Course Overview (In-person and Remote)

	CHEM 8A	**CHEM 8B**	**CHEM 8C**
Terms offered:	Fall and Winter	Winter and Spring	Spring and Fall Summer (5 week)
In-person Enrollment:	300-330 per section	270 per section	270 per section 100 (Summer)
Online Enrollment:	300-330 per section Fall 2020, Winter 2021	200-230 per section Spring 2020	200 Summer 2020
In-person Routine:	3 50-minute lectures for 10 weeks 8 flipped classroom activities replacing lectures	3 50-minute lectures for 10 weeks 7 flipped classroom activities replacing lectures	3 110-minute lectures for 5 weeks No flipped classroom activities
Online Routine:	3 synchronous 50-minute lectures for 10 weeks 8 flipped classroom activities	3 synchronous 50-minute lectures for 10 weeks No flipped classroom activities	3 synchronous 110-minute lectures for 5 weeks No flipped classroom activities
Course topics:	Bonding principles Acid-base properties Conformational analysis Stereochemistry Substitution reactions Elimination reactions Addition reactions Radical chemistry	Spectroscopy Conjugation Aromatic reactions Alcohols and ethers Carbonyl compounds Carboxylic acids	Carboxylic acid derivatives Alpha carbon chemistry Amines Amino acids Biomolecules Polymers

Table 2. Comparison of Traditional and Remote Learning Assessments

	Traditional In-Person Course	*Remote Learning with Activities*
Grading Basis	**85% Exams** *3 Exams in total* **15% Learning Modules** *Clickers* *Online Quizzes*	**35% Exams** *Final Exam only* **50% Homework** *8 Assignments in total* **15% Learning Modules** *Clickers* *Online Quizzes*
Assessment Type	85% High-Stakes Summative 15% Low-Stakes Formative	35% High-Stakes Summative 65% Low-Stakes Formative

Ultimately, summative assessments were significantly deemphasized with the transition to online instruction with a reduction from 85% of the final grade to a mere 35%. While this did not eliminate higher-stakes assessment within the course, it changed the overall mode of course assessment. The points that would have resulted from higher-stakes midterm exams were redistributed to weekly, lower-stakes online homework. Despite the benefits of transitioning to more formative assessment, summative assessment is necessary to ensure students have achieved a mastery level that ensures that they are adequately prepared for subsequent classes in the series. However, it is challenging to have meaningful summative assessment online, mainly when the most accessible online testing formats (e.g., multiple-choice) are typically authored for lower-order skills (3).

Low-Stakes Online Quizzing and Polling

One aspect of traditional instruction that is most easily transitioned into the online mode is day-to-day lecture content. This may be accomplished in two ways depending on other factors in the course: asynchronous video content or synchronous live lectures. In recent years, the amount of suitable educational content online has increased, and it may be possible to use existing resources to populate all course content areas. Not only this, but technology has progressed and become affordable to create individualized online content for students authored by individual instructors. In my modularly flipped classroom, I have generated video content for flipped classroom activity days using screen capture (e.g., Camtasia (15)) or lightboard (e.g., Learning Glass (16)) technology. While students appreciate this content's availability, it may suffer from some of the drawbacks of a traditional lecture. It will be mostly non-interactive and authoritative in presentation style. This content may be augmented with integrated online quizzing (e.g., PlayPosit (17)) or used in conjugation with online quizzing in an LMS; it will remain largely authoritative. However, paired with an in-class activity and discussion, a lesson with more interactive, dialogic characteristics may result (18). To avoid students becoming isolated with asynchronous content delivery, I elected to hold synchronous online lectures using Zoom and to integrate those with online clicker (e.g., PollEverywhere) questions. Students appear prefer online synchronous lecture formats as they have opportunities to engage with their instructor and each other with the added advantage of anonymity (19). Others have noted that engagement with online lectures were greatest when students had grade incentives to attend class regularly (20).

Clicker systems have found widespread use in higher education and are praised for their ability to engage students and provide valuable opportunities for formative assessment and immediate feedback on learning (*21*). My own experience integrating clickers into my online synchronous lectures involved using 3-5 questions per 50-minute lecture period. As in a traditional lecture, this would allow for chunking of material so that students could have ample time to process what they had just learned (*22*). They may also be used to measure student retention from previous lectures so that targeted review or just-in-time teaching could be used (*14*). As part of my classroom, clickers were a small portion of a student's grade, making up a part of their learning module scores. Students were also encouraged to participate by earning points for merely answering questions regardless of correctness. Lacking the grade incentive to simply be correct, students are more likely to use this as an opportunity to self-assess their learning. Students appreciate clicker polling as a way to anonymous answer questions and receive feedback on their learning (*23*). Using clickers for formative assessment during synchronous online lectures also has the advantage of encouraging student attendance in such classes.

In addition to traditional synchronous lectures online, my classroom typically also featured flipped modules administered on an approximately weekly basis. These flipped classes were structured with an assignment to watch videos and read book sections before class and a lecture tutorial activity during class. These videos and reading assignments were the primary mechanism for knowledge transfer; while the class periods were reserved for problem solving and sense making. While students commonly seek out online video explanations during their exam studying because they can control the pace and presentation of the video (*24*), they do not always see the necessity in preparing for an in-class activity. Online quizzes are paired with or integrated into the prescribed videos to ensure that students come to class prepared. In prior academic terms, integrated quizzing (i.e., Playposit) was used to ensure students were engaged during video viewing. PlayPosit quizzes are built around the existing video so that students are quizzed at predetermined time points. These in-video quizzes serve two purposes: one, it ensures students are watching the video and are introduced to the concepts before class activities; and two, it provides an opportunity for formative assessment and students to reflect upon the content beyond simply taking notes on what was presented. It is known that online video assigned as part of flipped classrooms is often not viewed as necessary by students and that using assessment as part of the video assignment increases compliance in viewing the posted videos (*25*). This can also be accomplished in other formats like LMS quiz or a short assignment in a homework system. Students must receive feedback on their quizzing as soon as possible to be most beneficial. Within the integrated video quizzes, students receive individualized feedback based on their answers as they progress through the video. An LMS quiz may be structured so that students receive immediate feedback in the form of a score or individualized problem-level feedback based on their answers. If a score is provided immediately, students may be allowed to retake the quiz to identify the errors they had made. If a score is not immediately offered, it would be beneficial for students to have quiz scores before working collaboratively on an in-class activity. Others have used free response questions and provided individualized feedback to students (*26*); heat however this may be limited to smaller class sizes where it does not represent an oversized time commitment for the instructor. Alternatively, this formative assessment might be reviewed by the instructor so that a short lecture might introduce the lecture tutorial activity and address common misconceptions identified in the online quiz (*14*). In my classroom, these online quizzes are incorporated into students' learning module grade and make up a small fraction of the available points; however, students appreciate their points towards their final grade for the effort they expend.

Student Impressions

In course evaluations, a student commented: *"I also really like the lecture polls, as it gives me a chance to actively engage in the material and tests me throughout the lecture; this is extremely useful because it is very easy for students like myself to doze off in the middle of Zoom lectures and having lecture poll questions about the content forces me to pay attention, which is something that all us students need."* Because a single day's polling is a small part of the students' grade and the course is structured such that some points can be missed without penalty, students can elect not to attend class if other obligations prevent them from doing so. This became incredibly important in the context of the pandemic-related shutdown as many students experienced job and family obligations that sometimes impeded their daily attendance and overall performance (27).

Active Learning in an Online Classroom

My classroom is structured in a partially flipped mode, where roughly every third lecture is taught as a flipped lecture tutorial (28). Initially transitioning to remote instruction due to pandemic-related shutdowns, my primary concern was keeping the classroom functional and not overwhelming students with a different schedule for each day of the week. The first goal was to implement an effective online learning environment focusing on the transition of the traditional lecture into the online classroom. Less effort was spent attempting to implement flipped classroom methods. As the first three weeks of CHEM 8B dealt predominantly with spectroscopy, that lent itself to problem-solving activities, which I typically would use for several class periods. Following lectures on topics including IR spectroscopy, mass spectrometry, and NMR spectroscopy, several lecture tutorials were used to allow students opportunities to practice and discuss this material. These lecture tutorials were distributed as PDF files, and students were to work collaboratively to solve these problems. Breakout rooms in video conferences were utilized; however, there are limitations to this. In a very large online class, technical limitations did not make it possible to have the same ideal size of student groups for lecture tutorial work. In-person, I typically instruct students to work in groups of 2-4 students each; the limitations in video conferencing software meant that I could have no more than approximately 40 breakout rooms for the large class, resulting in breakout rooms containing 8-12 students. While discussion could occur in such groups, it was usually dominated by 3-4 voices, while other students took a more passive role. In-person classroom activities also utilize learning assistants to answer clarifying questions, encourage discussion, and model thinking. Learning assistants are past students who have excelled in the course that have been trained to interact with students during activities (29, 30), but because they are former students experiencing the same stress related to their classes transitioning to remote learning, I elected not to use learning assistants during the spring term. Since the initial online term, learning assistants have been reincorporated into lecture tutorial standard operating procedures. Teaching assistants, learning assistants, and I would circulate between breakout rooms to answer any questions students might have. In the move to remote instruction, the quality and quantity of interactions with students have changed significantly. While most students elect to remain off-camera in these interactions, students extensively used chat to discuss the material, and interact with each other and the instructor using online whiteboard functionality. Often the lecture tutorial worksheet would feature annotations by multiple students working together to make sense and solve problems. These changes in interactions were noted by others who utilize learning assistants in their classrooms (31).

Two different methods of assessment and feedback were attempted in these collaborative activities. Firstly, web-based clickers were used to provide feedback on student progress through the activity, not necessarily by giving feedback on the correctness of their final answer, but rather by using intermediate points in the problem-solving process. By doing so, it would be possible for students to ask for clarification on their problem-solving and narrowing down what they might not understand or about what they have a misconception. This method is similar to the day-to-day use of clickers as formative assessments in normal lecture periods online. The second method of assessment and feedback was subsequently used where students input their answers into an online assignment after class. The online assignment was an automatically graded multiple-choice assessment that provided students rapid feedback after class on the correctness of their responses. Answer keys with full explanations were also provided so students could reflect on their work and see by what they might have been especially challenged.

Student Impressions

Students appreciated the formative assessment and feedback on their learning in the collaborative lecture tutorial sessions. A student from the fall quarter commented: *"The structure of having the first two classes of the week lectures and the last class of the week a group activity was helpful in confirming my understanding of the concepts and material learned each week."* Other students appreciated the regular nature of these activities that encouraged more frequent engagement with the material, commenting: *"The activities every week really helped me learn all the material without feeling like I was falling behind. I could see where I was at with the activity and ask questions on the parts I was confused with."* However, some students were frustrated by the limitations of the online environment and the unwillingness of their classmates to buy into the activity fully: *"It was difficult to get people to discuss the worksheets and work together. I might suggest asking people to turn on their cameras for this part if possible to get people to be more present for the activities."* Reflecting, these activities may be more beneficial as implemented in teaching assistant-led discussion sections where breakout rooms could be better utilized to ensure greater participation levels from all students.

Frequent Online Homework

Online homework became the cornerstone assessment used in my classroom in the transition to online instruction. Online homework sets for chemistry are often commercially available via various textbook publishers (e.g., ALEKS (32) and Sapling (33)) and open-source alternatives like OpenOChem (34). These homework websites have a wide variety of problem types available, predominantly multiple-choice, but some include free response type questions that involve free response chemical drawing. I had previously used online homework systems featuring both multiple-choice and chemical drawing options as formative assessments. However, the learning curve involved using chemical drawing software was a challenge for students, particularly when AI-assisted grading features were used. In recent years, I elected to provide optional homework for students to prepare them for summative assessments. These consisted of a combination of textbook problems as well as self-authored questions. Anecdotally, students strongly preferred the instructor-authored preparatory problems because of more substantial similarity to the exam tasks and style. Based on my experience and the rapid transition to online instruction, I elected to adapt previous problem sets as graded online homework.

Initially, these homework assignments were distributed via a shared file. Students could download their copy of this assignment to complete in whatever manner they chose. Some students printed blanks, scanned their completed work (often using cell phone apps to accomplish this) and submitted it as a PDF file. Other students with access to tablets simply annotated the homework file itself. Students commonly typed their answers into the document directly and used cell phone images of chemical structure drawings. Students were encouraged, but not required, to use chemical drawing software and were provided links to download Marvin (35); however, few students elected to do this. Students initially were resistant at using chemical drawing software because of the associated learning curve. In the fall quarter, the associated lab class instructed students on Marvin use, and a large segment of the class elected to use that software. To collect homework and provide grades and feedback, students uploaded their assignments via Gradescope. Gradescope is an online grading service that allows for pre-programmed rubrics to enable rapid, push-button grading (36). I used Gradescope for exam grading for the past three years but only recently implemented the homework feature. Despite being essentially "push-button" grading, rubric items can be detailed, and it enables students to see what types of answers receive credit. Grading is quick and efficient, so students receive feedback within a day or two after submission (when graded by 2-3 teaching assistants). Gradescope also contains premium features, such as automated grading for certain types of problems (fill-in-the-blank and formulas), which was quite useful for homework questions relating to spectroscopy, where formulas and numerical answers are sometimes used. Subsequent quarters of the class have used the "online assignment" feature within Gradescope, allowing the assignment to be built directly in Gradescope itself, not relying on a different website to distribute homework files. The online assignment option has several problems as options, including multiple-choice, multiple-answer, short answer, free response, and file upload options. Combined with Marvin, the file upload option enabled easy submission of mechanism problems commonly used in organic chemistry.

Homework assignments were inherently open-note, and Gradescope file uploads were used for most problems; this allowed for most questions to be free response and application-based in nature. Students could not merely look up the answer but instead were required to understand the material to a degree to apply it to solve a more complex problem. Each homework assignment was low-stakes (less than 5% of final grade) and meant to be formative. Because of this, students were encouraged to work through problems first on their own as a self-test and later with their notes and book. While in-person, students often formed informal study groups to study and discussion problem sets, however during remote instruction, students were no longer on campus and did not have access to their informal study groups; thus, student discussion was encouraged via the online class message board, Piazza (37). The online message board has been a useful tool in my classroom to allow students to ask questions and make their thinking visible by answering each other's questions. The classroom norm was not to ask specific questions nor share specific answers but to model thinking and ask clarifying questions.

Student Impressions

Students were largely supportive of the emphasis of homework, both for learning and as a determinant of their grade. In the end-of-course evaluations, one student commented, *"I believe having homework assignments instead of exams was the perfect choice to make and I hope other professors change to that. Although the homework would take a good chunk of time, it was worth it because it made sure that the student is on top of notes because you cannot do the homework without actually watching AND understanding what the professor is saying in his lectures."* Students recognized that the homework assignments encouraged frequent engagement with the material and that such a strategy led to superior learning outcomes, commenting: *"I found myself learning a lot more with weekly assignments, although they were lengthy, they truly assessed my understanding of the material. Having weekly assignments allows us to reflect and evaluate what we have learned during the week. This keeps us on track, and almost forces us to practice until we can grasp the material with the incentive that it contributes to our grade."*

Students noted the dual role of homework: first, to practice and self-assess; second, evaluate their progress throughout the quarter. In this way, homework itself served as a formative assessment with some limited summative assessment as well. Students appreciated the feedback on the homework as part of the learning process, commenting on the Gradescope homework system: *"I will say that Gradescope was ideal for turning in our assignments and receiving feedback. The rubric was helpful in knowing why points were lost."* Multiple-choice assessments often give students feedback that is too general; however, students could see what types of answers received credit with the rubrics.

Rethinking Online Exams

Even with the significant shift to formative assessment, summative assessment is necessary to ensure academic rigor within the class, particularly to determine if students were academically prepared to undertake the next course in the sequence. When my course was transitioned to online instruction, it was intentionally organized to require that adequate performance be needed on a final assessment to pass the class. Previous research has suggested that exams administered via computer and paper result in comparable test scores (*38*). Perhaps one of the most challenging aspects of shifting to online instruction was how to assess their organic chemistry knowledge fairly and effectively. Two common concerns with online assessment are: 1) ensuring academic integrity and preventing cheating (*4*); 2) multiple-choice assessments are the easiest to implement online, but often are not as academically rigorous because they lend themselves to lower-order or algorithmic problem-solving (*3*).

To address the first concern about academic integrity, several external proctoring services (e.g., ProctorU (*39*)) exist to meet this demand. While these services had been used by colleagues previously, with the rapid transition to online instruction worldwide, there was limited availability for new classes to adopt such technologies. The cost associated with such services is a concern, particularly with student populations with high rates of financial disadvantage. In response to these challenges, an in-house solution to online proctoring was developed using existing video conference software and teaching assistants. Other solutions exist to prevent academic dishonesty, including browser lockdown software and self-proctoring software (e.g., Yuja (*40*)) that records both screen and webcam for later review if cheating is suspected. I elected to use self-proctoring software for my class to discourage unauthorized collaboration between students. The final exam itself was offered as open-note and open-book in nature, so the ability and incentive to cheat are minimized. This approach may be applicable when the subject matter is applied as opposed to simply recall in nature. Evidence for cheating has been minimal, with no material identified on popular answer sharing websites or found on student chat servers. That is not to say that cheating has been non-existent, however it seems no more problematic than what is observed in person. Some have also made this observation about cheating in their courses (*41*); while others have noted that even with precautions such as test banks and ethics pledges, students were likely to seek out answers in an unethical manner (*42*).

The exam itself was administered as a multiple-choice exam using question banks in the LMS. The multiple-choice format was chosen to ensure that the exam could be graded quickly, free from grader bias and misapplication of a rubric. Using multiple-choice questions from a question bank also had the benefit of allowing for each student to receive a unique set of questions that would impede unauthorized collaboration during the exam. The challenge is offering a multiple-choice exam is in authoring questions that assess both lower-order and higher-order skills. Lower-order questions that require students to recall or describe are relatively easy for the instructor to generate and are easier for students to use test-taking strategies to arrive at the probable answer. As most instructors may already have existing free-response exam questions used in-person formats that test higher-order skills (i.e., Bloom levels of evaluate, analyze, or apply), it may be easiest to translate those questions into a multiple-choice question (43). For example, as part of first-quarter organic chemistry, students would typically be tasked with ranking organic molecules based on their predicted acidity (Figure 1a). Typically, students would be free to rank the molecules in any order, but in the multiple-choice format, the options are limited, with specific distractors and reasonable misconceptions included as answer options. Learning management systems often have other question options that may be used for such a question, such as ranking. Structuring this problem as a ranking question would not require the instructor consider common student misconceptions as part of authoring multiple choice answer options. Another common problem used in introductory organic chemistry is to predict the product of a given reaction (Figure 1b). Usually this problem would be open-ended, where students would have to draw the structure themselves with no suggested answer given. In this example, students are presented with possible structures with common distractors, including mechanism type, stereochemical outcome, and leaving group effects. One potential advantage of offering a problem in this format is that students are less likely to be distracted by unimportant factors but instead focus on the relevant information based on the possible answers provided, for example students in the past have previously attempted to incorporate DMSO and the sodium ion into the free response answer.

Direct, meaningful comparisons of exam performance in-person and online are not feasible due to the significant differences in exam format. Being multiple-choice in format, online exams lead students to guess at least to score any points, therefore students are likely to have a higher baseline exam score online. In-person assessments on the other hand are largely free response in nature and students cannot easily guess to earn points by chance, therefore making the baseline for these assessments closer to zero. Exam scores are indeed higher with online exams than with in-person exams. This fact and combined with a modified final grade scale results in similar grade distributions for both online and in-person classes. The sequence and circumstance of the courses taught online has limited the ability generalize on student performance. While CHEM 8B and 8C were taught in sequentially, only a small fraction of students elected to take a summer course as is typical for summer course enrollment. These students were successful as other cohorts had been, but this may be due to self-selection.

Student Impressions

Student impressions of exams were harder to gauge by course evaluations. Students were largely satisfied with the transition from summative assessments like mid-term exams to formative assessments like homework, one student commenting: *"I really enjoyed the substitution of midterms with homework. This was my second time taking Chem 8B, so being able to sit down and thoroughly delve through each exam-like problem was really useful to be able to understand the material. I feel much more prepared, this time around, for the final."* (For other student impressions, see the Online Homework section.) However, some students were uncertain about their upcoming final exam. One student commented: *"I'm stressed for the final because we had no midterms so I wish he would bring those back to prepare me."* A subset of students did appear to value the infrequent final exams; those who did seemed to struggle with the time demands of more frequent homework assignments. Students may have been under the impression that homework problems were not adequate preparation for the final exam or that the questions would be in a significantly different format. With most students supporting this new grading scheme and overall student success, I anticipate that features of these exams and grading schemes will find their way into traditional lecture classrooms.

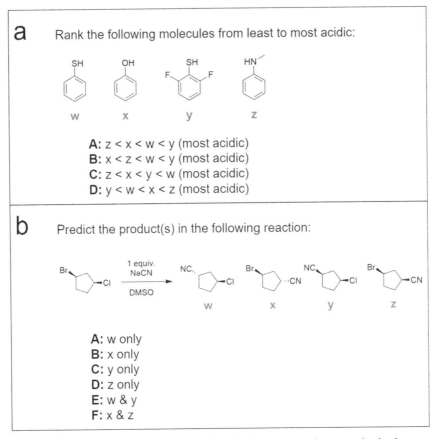

Figure 1. a) Ranking-based multiple-choice problem. b) Reaction prediction multiple-choice problem.

Applying Lessons Learned Towards Future Instruction

Overall, the emergency transition from large-format live lectures to remote instruction was well received by students. Anecdotal comments from students stated they preferred the lower-stakes and more frequent formative assessment, especially when many other courses did not change significantly when transitioned to remote learning. Some students stated they missed the typical in-class activities, including lecture tutorials. It was possible to leverage technology to facilitate a richer learning environment, but additional personnel were required to implement this effectively. Students have rapidly adapted to remote instruction requirements, and the overwhelming majority were not impeded using new technology. The assessment modifications were a welcome change to the course and improved the course quality according to students.

Looking to future courses online, teaching practices and assessment can continue to evolve and be refined. While these past few terms have been a learning experience for both instructors and students, I am confident that students can adapt to more demanding online implementations of active learning. The more frequent, lower-stakes assessments are highly beneficial to student learning. It is worth exploring other implementations of such, keeping in mind both academic rigor and flexibility for our students.

Previously, I would not have considered online homework feasible for such a large in-person class due to limitations in grading and providing meaningful feedback; however, online homework can be implemented relatively quickly and easily with significant benefit for students. I anticipate this will become standard operating procedure in the future in-person classes, in addition to online classes.

In addition to assessing our students for purposes of demonstrating mastery, it is essential that instructors also assess how successful their class is for their students. Instructors should keep communication lines open so that students can provide feedback on our instruction; for example, an informal mid-term survey can provide valuable feedback to improve our teaching practice. This feedback not only can be used to improve our courses but also can be shared with the wider chemistry teaching community to enhance teaching in the field more generally.

References

1. Bozkurt, A.; Sharma, R. C. Emergency Remote Teaching in a Time of Global Crisis Due to CoronaVirus Pandemic. *Asian J. Distance Educ.* **2020**, *15* (1), 2020. https://doi.org/10.5281/zenodo.3778083.
2. A. Holme, T. Introduction to the Journal of Chemical Education Special Issue on Insights Gained While Teaching Chemistry in the Time of COVID-19. *J. Chem. Educ.* **2020**, *97* (9), 2375–2377. https://doi.org/10.1021/acs.jchemed.0c01087.
3. Hartman, J. R.; Lin, S. Analysis of Student Performance on Multiple-Choice Questions in General Chemistry. *J. Chem. Educ* **2011**, *88*, 1223–1230. https://doi.org/10.1021/ed100133v.
4. Greaser, J.; Black, E. W.; Dawson, K. Academic Dishonesty in Traditional and Online Classrooms: Does the "Media Equation" Hold True? *Online Learn.* **2008**, *12* (3), 23–30. https://doi.org/10.24059/olj.v12i3.13.
5. P. Dicks, A.; Morra, B.; B. Quinlan, K. Lessons Learned from the COVID-19 Crisis: Adjusting Assessment Approaches within Introductory Organic Courses. *J. Chem. Educ.* **2020**, *97* (9), 3406–3412. https://doi.org/10.1021/acs.jchemed.0c00529.
6. Shute, V. J. Focus on Formative Feedback. *Rev. Educ. Res.* **2008**, *78* (1), 153–189. https://doi.org/10.3102/0034654307313795.
7. Gaytan, J.; McEwen, B. C. Effective Online Instructional and Assessment Strategies. *Int. J. Phytoremediation* **2007**, *21* (1), 117–132. https://doi.org/10.1080/08923640701341653.
8. Hondrich, A. L.; Decristan, J.; Hertel, S.; Klieme, E. Formative Assessment and Intrinsic Motivation: The Mediating Role of Perceived Competence. *Zeitschrift fur Erziehungswiss.* **2018**, *21* (4), 717–734. https://doi.org/10.1007/s11618-018-0833-z.
9. Angus, S. D.; Watson, J. Does Regular Online Testing Enhance Student Learning in the Numerical Sciences? Robust Evidence from a Large Data Set. *Br. J. Educ. Technol.* **2009**, *40* (2), 255–272. https://doi.org/10.1111/j.1467-8535.2008.00916.x.
10. Richards-Babb, M.; Curtis, R.; Georgieva, Z.; Penn, J. H. Student Perceptions of Online Homework Use for Formative Assessment of Learning in Organic Chemistry. *J. Chem. Educ.* **2015**, *92* (11), 1813–1819. https://doi.org/10.1021/acs.jchemed.5b00294.
11. *Host interactive online meetings | Poll Everywhere | Poll Everywhere*. https://www.polleverywhere.com/ (accessed Jun. 17, 2020).
12. Skinner, S. On Clickers, Questions, and Learning. *J. Coll. Sci. Teach.* **2009**, *38* (4), 20.

13. Canelas, D. A.; Hill, J. L.; Carden, R. G. *Cooperative Learning in Large Sections of Organic Chemistry: Transitioning to POGIL*; UTC: 2019. https://doi.org/10.1021/bk-2019-1336.ch012.
14. Mazur, E.; Watkins, J. Just-in-Time Teaching and Peer Instruction. In *Just in Time Teaching: Across the Disciplines, and Across the Academy*; 2009; pp 39–62.
15. *Camtasia: Screen Recorder & Video Editor (Free Trial) | TechSmith*. https://www.techsmith.com/video-editor.html (accessed Mar. 8, 2021).
16. *Home - Learning Glass*. https://www.learning.glass/ (accessed Mar. 8, 2021).
17. *PlayPosit | Interactive Video Platform*. https://go.playposit.com/ (accessed Mar. 8, 2021).
18. Scott, P. H.; Mortimer, E. F.; Aguiar, O. G. The Tension between Authoritative and Dialogic Discourse: A Fundamental Characteristic of Meaning Making Interactions in High School. *Science Education* **2006**, 605–631. https://doi.org/10.1002/sce.20131.
19. Young, S.; Young, H.; Cartwright, A. Does Lecture Format Matter? Exploring Student Preferences in Higher Education. *J. Perspect. Appl. Acad. Pract.* **2020**, *8* (1), 30–40. https://doi.org/10.14297/jpaap.v8i1.406.
20. A. Perets, E.; Chabeda, D.; Z. Gong, A.; Huang, X.; Sang Fung, T.; Yi Ng, K.; Bathgate, M.; C. Y. Yan, E. Impact of the Emergency Transition to Remote Teaching on Student Engagement in a Non-STEM Undergraduate Chemistry Course in the Time of COVID-19. *J. Chem. Educ.* **2020**, *97* (9), 2439–2447. https://doi.org/10.1021/acs.jchemed.0c00879.
21. Fies, C.; Marshall, J. Classroom Response Systems: A Review of the Literature. *J. Sci. Educ. Technol.* **2006**, *15* (1), 101–109. https://doi.org/10.1007/s10956-006-0360-1.
22. Caldwell, J. Clickers in the Large Classroom: Current Research and Best-Practice Tips. *CBE - Life Sci. Educ.* **2007**, *6*, 9–20. https://doi.org/10.1187/cbe.06-12-0205.
23. Wijtmans, M.; van Rens, L.; E. van Muijlwijk-Koezen, J. Activating Students' Interest and Participation in Lectures and Practical Courses Using Their Electronic Devices. *J. Chem. Educ.* **2014**, *91* (11), 1830–1837. https://doi.org/10.1021/ed500148r.
24. Casselman, M. D.; Atit, K.; Henbest, G.; Guregyan, C.; Mortezaei, K.; Eichler, J. F. Dissecting the Flipped Classroom: Using a Randomized Controlled Trial Experiment to Determine When Student Learning Occurs. *J. Chem. Educ.* **2020**, *97* (1), 27–35. https://doi.org/10.1021/acs.jchemed.9b00767.
25. He, W.; Holton, A.; Farkas, G.; Warschauer, M. The Effects of Flipped Instruction on Out-of-Class Study Time, Exam Performance, and Student Perceptions. *Learn. Instr.* **2016**, *45*, 61–71. https://doi.org/10.1016/j.learninstruc.2016.07.001.
26. Dimick Gray, S. Embedded Video Questions as a Low-Stakes Assignment During the Remote Learning Transition. *J. Chem. Educ.* **2020**, *97* (9), 3172–3175. https://doi.org/10.1021/acs.jchemed.0c00505.
27. Jones, H. E.; Manze, M.; Ngo, V.; Lamberson, P.; Freudenberg, N. The Impact of the COVID-19 Pandemic on College Students' Health and Financial Stability in New York City: Findings from a Population-Based Sample of City University of New York (CUNY) Students. *J. Urban Heal.* **2021**, *98* (2), 187–196. https://doi.org/10.1007/s11524-020-00506-x.
28. Casselman, M. D. Effective Implementations of a Partially Flipped Classroom for Large-Enrollment Organic Chemistry Courses. *ACS Symp. Ser.* **2019**, *1336*, 187–196. https://doi.org/10.1021/bk-2019-1336.ch011.

29. Sellami, N.; Shaked, S.; Laski, F. A.; Eagan, K. M.; Sanders, E. R. Implementation of a Learning Assistant Program Improves Student Performance on Higher-Order Assessments. *CBE—Life Sci. Educ.* **2017**, *16* (4), ar62. https://doi.org/10.1187/cbe.16-12-0341.

30. Repice, M. D.; Sawyer, R. K.; Hogrebe, M. C.; Brown, P. L.; Luesse, S. B.; Gealy, D. J.; Frey, R. F. Talking through the Problems: A Study of Discourse in Peer-Led Small Groups. *Chem. Educ. Res. Pr.* **2016**, *17* (17), 555–568. https://doi.org/10.1039/c5rp00154d.

31. Emenike, M. E.; Schick, C. P.; Van Duzor, A. G.; Sabella, M. S.; Hendrickson, S. M.; Langdon, L. S. Leveraging Undergraduate Learning Assistants to Engage Students during Remote Instruction: Strategies and Lessons Learned from Four Institutions. *J. Chem. Educ.* **2020**, *97* (9), 2502–2511. https://doi.org/10.1021/acs.jchemed.0c00779.

32. *ALEKS – Adaptive Learning & Assessment for Math, Chemistry, Statistics & More*. https://www.aleks.com/ (accessed Mar. 8, 2021).

33. *Sapling Learing and SaplingPlus*. https://www.macmillanlearning.com/college/us/digital/sapling (accessed Mar. 8, 2021).

34. *OpenOChem Cloud | OpenOChem*. http://www.openochem.org/ooc/ (accessed Mar. 8, 2021).

35. *Marvin | ChemAxon* https://chemaxon.com/products/marvin (accessed Jun. 17, 2020).

36. *Gradescope*. https://www.gradescope.com/ (accessed Jun. 17, 2020).

37. *Piazza • Ask. Answer. Explore. Whenever*. https://piazza.com/ (accessed Jun. 17, 2020).

38. Prisacari, A. A.; Holme, T. A.; Danielson, J. Comparing Student Performance Using Computer and Paper-Based Tests: Results from Two Studies in General Chemistry. *J. Chem. Educ.* **2017**, *94* (12), 1822–1830. https://doi.org/10.1021/acs.jchemed.7b00274.

39. *ProctorU - The Leading Proctoring Solution for Online Exams*. https://www.proctoru.com/ (accessed Mar. 1, 2021).

40. *Video Test Proctoring - YuJa Official Home Page*. https://www.yuja.com/video-test-proctoring/ (accessed Mar. 8, 2021).

41. O'Carroll, I. P.; Buck, M. R.; Durkin, D. P.; Farrell, W. S. With Anchors Aweigh, Synchronous Instruction Preferred by Naval Academy Instructors in Small Undergraduate Chemistry Classes. *J. Chem. Educ.* **2020**, *97* (9), 2383–2388. https://doi.org/10.1021/acs.jchemed.0c00710.

42. Clark, T. M.; Callam, C. S.; Paul, N. M.; Stoltzfus, M. W.; Turner, D. Testing in the Time of COVID-19: A Sudden Transition to Unproctored Online Exams. *J. Chem. Educ.* **2020**, *97* (9), 3413–3417. https://doi.org/10.1021/acs.jchemed.0c00546.

43. Dirks, C.; Wenderoth, M.; Withers, M. Assessing Higher-Order Cognitive Skills with Multiple-Choice Questions. In *Assessment in the College Science Classroom*; Blake, S., Ed.; W.H. Freeman and Company: New York, 2014; pp 74–91.

Chapter 3

A First Semester General Chemistry Flipped Remote Classroom: Advantages and Disadvantages

Wendy E. Schatzberg*

Department of Chemistry, Dixie State University, Saint George, Utah 84770, United States
*Email: schatzberg@dixie.edu

Due to the Covid-19 pandemic in the fall 2020 semester, a traditional in-person class was redesigned into an online remote flipped classroom. Students' average grades, American Chemical Society exams, and a metacognition survey were given to compare the remote flipped class to the traditional in-person classes. The comparison was done to have a general assessment of the effectiveness of the remote flipped classroom. Students had a slightly higher average grade, exam score, and elaborated more on the metacognition survey. This is seen as an indication that the redesign was successful for the semester and further research is being done to see if these findings were an outlier or an overall trend.

Introduction

Remote teaching flipped classrooms forces students to work at home via the internet rather than a traditional in-person lecture (1). Watching the required lecture videos and doing the required worksheet replaced traditional lectures. While students could decide to skip either the video lecture or class, the required worksheets associated with lecture video and in-class Zoom meetings incentive for students to attend and do the work. The lecture videos and synchronous Zoom meetings supported students working through problems, students could decide to attempt to do the worksheets without support and alone but students reported that it was easier to do them with the class structure. The flipped classroom method (2) suits itself well to remote and pandemic teaching where the unknowns in teaching methodology and semester setup mean that the professor has to be extremely flexible in teaching delivery. Hyflex, remote, in-person, etc. all can be done with the flipped classroom and be transitioned between methodologies easily since most of the lecture content is on the online educational platform (3). The flipped classroom also provides a lot of structure to student learning both inside and outside the classroom, wherever the classroom may be. This structure in a time of unstructured and uncertain life could keep a student on track in their studies no matter how their university or personal life changes. They always have built-in lecture and problem-solving (4). There is never a lack of accessibility if they have a computer and internet.

Online teaching has pros and cons – the easy availability of content can improve accessibility but can invite cheating. The asynchronous and synchronous online class schedule allows students flexibility with their schedules but can cause students who are ill-prepared time management skills to fall behind (5). Online can be asynchronous, synchronous, or a hybrid between the two but most commonly online courses are asynchronous, students work independently with minimal synchronous professor interactions. For this study, we will be referring to a course that contains all content available online (videos, worksheets, exams, etc.) but mandatory one-hour class synchronous meetings each day, four days a week (similar to the traditional in-person course), to be labeled a 'remote course'.

A remote teaching curriculum has been found to allow faculty to adjust quickly to different student needs (6). The Covid-19 pandemic required a teaching environment that could adapt quickly depending on the student and professor's health, safety, and university requirements. With this in mind, a flipped classroom seemed appropriate: it allowed for in-person, Hyflex, and/or remote to occur with minimal changes to student assignments. The remote lecture consisted of recorded professor lecture videos uploaded in Canvas; class time synchronous Zoom meetings were devoted to group work, problem-solving, and student issues as needed. If a student was absent due to quarantine and/or illness, the lectures were still available to the student and they were still able to access the class. Class time problem solving Zoom meetings were also recorded for students to watch later if needed.

There have been many studies about flipped classrooms in the past decade. There appears to be a correlation of flipped classrooms with student success, critical thinking, and student satisfaction (7). There have also been studies showing that teaching a flipped classroom needs student buy-in from the beginning and professor support (7). Professors designing and implementing a flipped classroom need to redesign their curricula and that takes a lot of time and effort. Guidance for the student occurs not only in the classroom but outside since the lecture content is typically received outside of the classroom and inside the classroom. Students do need to have more time management skills and independence than found with traditional learners, and it has been postulated that the flipped classroom might put lower-achieving students at a disadvantage since they usually do not have the skill set that the higher achieving students do. Students that do not have time management skills can often fall behind in the class (8). The flipped classroom is also a different way of teaching than the traditional podium lecture which may increase student dissatisfaction with an unfamiliar teaching style. Another student complaint is that students may believe that they are teaching themselves rather than the instructor teaching them, due to a large amount of guided material outside of the classroom. Many have been found to have negative associations with this curriculum style.

A flipped classroom is different from a traditional classroom in that the student is spending time in the classroom doing homework and practice problems with the outside the classroom work is the traditional lecture. Students who have bad study habits of attending lectures but doing minimal work outside class meeting times find that the flipped classroom to be too much time due to them mentally allotting no time outside of the class for their work (9). These students tend to find the beginning of the flipped classroom semester to be the toughest because they were not mentally prepared to do work inside and outside the classroom. These students are probably the best students to receive the flipped classroom approach; it shows them and forces them to allot time inside and outside the classroom to learn the material and how much time they should be spending on their classwork. It is one thing to hear that they should be spending two hours a night on a class; it is another to be forced to do it.

Promoting metacognition skills in students, especially freshmen, has been a resource of much research. Previous studies have shown that survey prompts can use students' reflections before and after exams to encourage students to think about their preparation, performance, before and after exams (10). Students have found these reflections helpful at improving their exam outcomes and text anxiety as noted by the lecture professor. Previous semester studies have shown that students have increased articulation on the study habits, understanding, and test-taking ability after doing these reflections than with a control group. The act of writing and thinking about performance may be a neutral feeling about an event, a test that might have gone negatively emotionally. By creating a neutral environment, students may feel more comfortable analyzing the situation as an objective observer rather than an emotional participant and withdraw.

Methods

This first-semester general chemistry course was restructured from an in-person class to a flipped classroom; lecture content was provided in a recorded video found on the class software platform Canvas the day before a synchronous internet Zoom lecture. The previous in-person course met five days a week for 50 minutes each in a classroom with the only online component being grades and chapter homework. Those were also incorporated in the remote flipped course but there was vastly more online material than those two components, which will be further described.

The flipped remote classroom had a similar content exam (though the exams were now online rather than paper and in-person) and homework assignments. There was big difference between the two in terms of grading and points were the flipped remote classroom assignments. There were two additional assignments in the remote class, one associated with the video lecture and one associated with the class time. These were worth five points each and were meant to keep the students on-task in the remote teaching environment. In the remote flipped classroom there was a five-question graded worksheet required covering the flipped lecture material and a five-question graded question worksheet done in class. The worksheets were to incentivized students into watching the lecture video and attending the synchronous class. The grade percentages were the same as in previous semesters.

Exams in the course were all online, 50 minutes allotted, 20 multiple choice questions based on comprehensive material. There was an online proctoring system to deter students from cheating, Proctorio. There were a total of four exams, the lowest grade dropped from their course; this setup was similar to previous semesters. The final was the American Chemical Society (ACS) General Chemistry I final exam that was released specifically for remote teaching in Fall 2020. ACS exams are not typically allowed to be used online but due to the pandemic, the Exam Institute released a standardized exam for this purpose with the caveat that the exam was required to be deleted once the final was administered.

The class time was done via the computer software Zoom since the university had a license for all students and faculty. There were approximately 60 students enrolled in a first-semester general chemistry course that was notified that it was fully remote via Zoom two weeks before classes commenced. This announcement was to give students time to adjust if they disliked that venue. Canvas was the internet platform used by the study university and Zoom links were accessible through Canvas for class and office hours.

The students seem to drop out of the course and self-select out (4 of 60 students). This was a typical drop rate for this course. Students seemed to quickly learn that it was easier to watch the videos and participate in class than try to attempt to go asynchronous, attempting to use Google to do the assignments. Students did have to show their work on all worksheet problems; even if students were passively copying material done in synchronous lecture, the writing out the problem solving would at a minimum allow students to improve their learning.

At the beginning of the course, students were mostly unfamiliar with a flipped classroom, as reported by the professor. After the first week, students remarked that they liked the guidance of the videos, problem-solving, and independence. The majority of students in this course were first semester freshmen (80%) with the rest of the student population being a variety of sophomores, juniors, and seniors. Freshmen typically have been documented to have issues with retention due to unknown academic cultural norms and lack of time management. Students also may never have reflected upon their performance and accepted grades as a static event.

The metacognition surveys were given before and after each exam via the internet software Surveymonkey. Student names and answers were kept separate to keep the grading professor unaware what the students were answering but knew which students did answer the surveys to award extra credit. These questions asked the students how they were planning on preparing for an exam, what they thought was the most important thing to do before an exam, how they participated in class, how they were going to change their behavior after the exam, and what changes they were going to make for the next exam. These surveys were designed to be neutral prompt to promote students into metacognition; thinking about what methods work and doesn't work for them in the class and on assessments. It has been found that students often have not reflected upon their academic performance and have mostly considered outside sources (teachers, class type, etc.) as reasons why they are not performing at the level they wished. The prompt was to promote students to actively think about their actions and how they can tie them to performance. Previous research studies have shown that these types of prompts can prompt students to discover what variables in their life and study habits work for them in both the chemistry course and in their overall academic career (11). In regards to this study, surveys were analyzed for keywords, response length, and compared to semesters that were fully in-person in a traditional classroom.

Results

The exam grade average overall was 76% for the in-person class and 79% for the flipped remote course. Both classes had three 20 question multiple-choice exams throughout the semester with the fourth exam being the last day of class and comprehensive. They were allowed to drop the lowest exam from the semester. The final exam was the first semester American Chemical Society general chemistry exam with the average for in-person being 39 and the remote being 43 out of 70 questions. Exams for the flipped remote classroom were via the university platform Canvas with Proctorio monitoring the class for online cheating. Normally ACS exams cannot be used via the internet but a special allowance and exam were given due to the Covid-19 pandemic.

The average student grade in the remote flipped course, Fall 2020, class was 84.5% corresponding to a B grade. Compared to previous semesters of this course with the same professor (Table 1) this grade average is slightly higher than other semester averages. The course content was the same all four years and the average on the ACS first semester general chemistry final was similar though slightly higher with the Fall 2020 flipped remote class. While the ACS online exam was different than ACS exam in 2018 and 2019, the exam institute released information that the over exam averages was anticipated to be similar. As of this study, the 2020 exam average has not been released. The ACS first semester general chemistry norm is 39 with a standard deviation of 12.4, which students were performing at the norm in all reported years. It should also be noted that in this chemistry department, all chemistry courses for the undergraduate chemistry degree used ACS exams as final exams and this course was the only course that showed score improvement compared to previous years.

Table 1. Average student first semester general chemistry grades

	Fall 2018	Fall 2019	Fall 2020
Class Type	In-Person	In-Person	Remote Flipped
Class Grade Average	81.1, 78.3	73.5, 73.8	84.5
Grade	B-, C+	C, C	B
ACS Exam Average	39	39, 39.3	43
ACS Exam Median	38	38	43

Students took a metacognition survey before, open 7 days before an exam, and after their course exams, open for one week after an exam, in both the previous in-person traditional curriculum and with the remote flipped classroom. The timing of the surveys was kept the same to assist in being able to compare the two student populations. These surveys consisted of questions (Table 2) that asked students about their study habits, goals, and to reflect upon their previous performance. The average time for the student to answer the survey was three to five minutes. These types of surveys are a metacognition prompt (12) and assist students in evaluating their study skills. Students in both the in-person and remote flipped classroom gave more elaborate responses as the semester continued, this may be due to better reflection on their performance with more exams taken or familiarity with the survey questions. Students responded more to the surveys (Table 3) though that may be due to being more familiar and at ease with doing online work in comparison to the in-person classes.

Table 2. Sample Metacognition Reflection Questions

Name three reasons why you were successful or unsuccessful on Exam 4.
How did you study for Exam 4?
What are you going to do the same preparing for the Final Exam?
What are you going to do differently to prepare for the Final Exam?

Table 3. Metacognition Response Rates

Survey Student Response Percentages	Fall 2018	Fall 2019	Fall 2020
	In-Person	In-Person	Remote Flipped
Survey 1, 2 (Exam 1)	35% / 40%	40% / 55%	65% / 65%
Survey 3, 4 (Exam 2)	50% / 40%	50% / 60%	85% / 80%
Survey 5, 6 (Exam 3)	50% / 60%	75% / 40%	75% / 75%
Survey 7, 8 (Exam 4)	55% / 55%	40% / 55%	50% / 50%

Considering that students were under a higher than normal stress situation in both their personal, professional, and educational lives because of the Covid-19 pandemic, achieving similar outcomes as a regular semester was considered a success. Many other classes at this university and in the department were recording lower than usual average exam and overall grades, with those courses using the Hyflex or remote teaching methodology. It is considered that this class due to being a single data point could be a statistical outlier and not a true representation of a flipped remote classroom success. Further work using other classes will continue in ongoing classes using flipped classrooms with the Hyflex model and with different student populations.

Attendance was a problem that may be due to the remote nature of the class. According to the class professor students would often not have on their computer cameras; this caused issues on whether or not the students were attentive to class. In the Zoom class meetings, students would be placed into Zoom breakout rooms, groups have randomly chosen of 5-6 students, and expected to work together to solve the problems. After approximately ten minutes, the professor would then bring all the students back to one main Zoom presentation room and call upon a random student to answer a problem or partial answer to a problem. If a student was not paying attention and/or absent it would be noticeable because the student wouldn't go into the breakout room and/or wouldn't have any answers to being called upon. There have been studies were professors have forced attendance via having internet cameras on at all times, there is controversy over students not wanting to disclose where they are at and what their home life is like (13–15).

Adjusting to a flipped classroom via remote created a different workload than normal. The compared in-person class had been taught that way for numerous years by the instructor. The summer before the fall semester required intense study of flipped classroom methodology and the online classroom curriculum. It was decided to create 10-15 minute classroom videos for each day with worksheets for the video due before the remote class that day. Students needed to upload the file before attending class to receive the points. The in-class remote was students going through another worksheet via the internet software Zoom in breakout rooms and students participating in class by the groups giving answers when prompted. It would not be suggested to do this way in the future, the grading load for the worksheets (2 per class), automatic grading of homework for each chapter, grading exams, the extra credit from the surveys, daily remote class, and recording videos/editing created a large workload burden. It would be recommended that the flipped classroom provided flexibility for the classroom in a time where it was uncertain if the university would be in-person, Hyflex, remote, or some other format. Having these classroom lecture videos, allowing for the instructor to be prepared for all contingencies and have a foundation to build a curriculum around. It would be suggested that flipped classroom videos would be a good idea to create for all classrooms – it could provide supplementary material for students attending class, absent, ill, etc. It should also be noted that some of the students attempted to skip the remote class and just watch the video at a later date. These students were consistently in the C, D, F, and W grades. Whether their grades were a result of the later video watching or just their basic knowledge level, it is suggested that videos must be either used with a graded worksheet to provide an incentive for the student to be active learning or as completely supplementary material to an in-person lecture.

Study Limitations

The control study of in-person students from previous fall semester general chemistry first semester classrooms consisted of three 60 person classes. The flipped remote online courses had one 60 person classroom. There could be differences in the data due to the smaller flipped remote classroom student population than with the comparison. Also, there were another two classrooms at the university that were in-person but were taught by different instructors. Those students were not used as a control due to the instructors' unwillingness to participate in the study due to their high workload and stress in the pandemic situation. Students may come from a different self-selecting group; the students wanting to take an online remote course may be a different population than students in a non-pandemic wanting an in-person class.

Flipped classrooms have been successful in other classrooms but there is limited data on doing it in a remote synchronous classroom environment. In this case study, I would say that there was a limited success with students and with instructor considerations. Students seem successful in the course and showed similar levels of metacognition to the in-person students. The instructor had the flexibility to handle an uncertain teaching situation that could change at any time. The high level of workload before the semester preparing lecture videos, online homework, lecture worksheets, and synchronous class worksheets, and online exams is a downfall to this approach. In further semesters, there will be an embedded question in the lecture for students to turn in at the end of the week, and a heavier usage of automatically graded online homework rather than depending on the instructor to grade two worksheets per day. If an instructor has time to plan, a flipped-classroom approach gives curriculum flexibility to an instructor and allows the instructor to decide if they wish to change their class to asynchronous online, synchronous remote, Hyflex, or in-person.

Technology accessibility was a possible downside to this study. Students needed to have a working computer and stable internet access. Due to the Covid-19 pandemic, universities did make technology available on campus and it was made open to students this academic year the technology requirements. Still, there were times that students had internet issues that interfered with class and had to be accounted for. The technology did allow for students to access the class no matter where they were giving higher accessibility than normal. This higher accessibility did have a downside according to the study professor; students reported in class that they were not treating the video lecture as if it was a true lecture. The students did not take notes, pay attention, and were passively learning. As an example, students in this study would schedule work shifts at the same time as their lecture, assuming that they could passively attend their classwork while actively doing other tasks. Having assignments due that needed to be worked on in class and student groups, these students quickly found that not participating gave them a reputation in class as a person no one wanted to work with. The professor would call upon randomly chosen students to answer questions and if the student was not paying attention and/or didn't attempt the problem, they would lose attendance points. It was using cultural norms in an attempt to keep students accountable for their learning and activities in an online setting.

Students may have had the habit of being passive learners using the internet because scrolling through websites, listening to podcasts, watching YouTube videos, which are all passive learning scenarios. Since the class was using the same internet as their previous passive learning experiences, the students transferred their previous habits into the class assuming they would have a good outcome. The students in classes need to change these passive learning student habits, but as with changing any habit, it takes time, effort, and mental flexibility. As long as a professor is open with their expectations and designs their course to take these variables into account, students may change to a more active learner mindset.

Discussion

The Covid-19 pandemic made the changes from a traditional in- person classroom to a flipped remote online teaching experience possible. The university was not supportive of online learning in the sciences or flipped classrooms. The flipped classroom approach allowed for the flexibility in curricula surrounding the uncertainty in fall 2020 instruction. The flipped classroom teaching was not well supported before the pandemic but since all instruction was considered unique, this seemed like a good time to try. The curriculum changes due to the pandemic allowed for alterations from a traditional in-classroom class to be adjusted to a flipped remote classroom model. As this was the first time the instructor had permission to do a flipped classroom, there could be less achievement due to that lack of experience. Students and instructors had higher levels of stress due to the pandemic that could have affected the study results.

Before an online exam, students had a practice quiz involving the online learning platform Canvas and the anti-cheating software Proctorio. This was done to allow students to familiarize themselves with taking an online exam and to work out any problems with their computers and browsers. Students had one day to complete the timed 50-minute, 20 multiple choice question exams. One student was caught cheating on the exam via Proctorio; this student actively walked away from the exam and hence was not allowed to continue. The student was reported to the student academic council and policy for cheating. Besides that one data point, the rest of the students did not have any issues and seemed to be treating the exams similar to an in-person exam.

Higher course grades in the flipped remote classroom could be attributed to numerous variables; student population, more guided instructor practices in lecture, students need to come to lecture prepared, more points due to the worksheets, etc. Further study needs to be done to eliminate variables and/or highlight which aspects caused the higher grades. This preliminary finding is a starting point for further research and should not be considered to give conclusive results. Higher grades also do not necessarily correlate with greater student understanding, while it is hopeful that students who participated in the study did learn more using this curriculum that cannot be definitively stated.

Students did not respond highly to the last survey, this is probably due to the timing of the survey where the announcement for the survey was done during Thanksgiving break and it was during the last full week of school. Students probably forgot or were distracted by so many end-of-semester activities that they did not complete the survey. The amount of respondents for the last survey is nearly half as much as other surveys; this survey was corresponding to the last exam in the class.

Student metacognition survey responses appear to be slightly more elaborate overall than with semesters that were fully in-person. Students wrote on average two sentences more while answering the survey questions. Students in the remote flipped classroom said that they referenced the online textbook more than the in-person classroom. This could be due to the remote nature of the class made students more aware of their textbook within the Canvas portal they used to access lectures. Remote students were used to using the online material that could comfortable communicating via that method as opposed to a pure in-person classroom. In-person classes had only one assignment online; each of the ten chapters' questions had an associated online homework problem set. The remote flipped class had two daily online assignments, along with daily online lecture videos, Zoom class meetings, monthly online tests, and supplementary online reference resources.

Students remarked in metacognition surveys how they liked the flipped classroom approach because they felt freer to come to their professor with their questions in problem-solving. That having lectured and taking notes outside of class, more passive student role, was preferred, and having the active problem solving be guided with fellow students and the professor was easier on them. That if they were having issues with parts of the problems or a concept that they knew they could come to class and have plenty of time to have those issues addressed. That the traditional classroom did not allow time for students to digest the material and practice it enough for the students to feel confident as to where they were having issues or where they had questions. That it took time for them to recognize that they did not understand a concept or method, something that traditional lecture does not give for students. By the time students understand that they do not completely understand the concept, they are alone without expert guidance. In that case, they usually relied upon the internet for guidance which may or may not have the correct guidance.

Students remarked in metacognition surveys that the daily assignments on the lecture video were an incentive for them to work on a class daily rather than pushing off studying and homework until the last minute. The video assignments consisted of five questions all relevant to the lecture and specifically designed not to be questions that could be answered by an internet search. Students did occasionally answer the questions using Wikipedia and other internet resources, as stated by the professor, which caused them to lose points and receive an email from the professor about watching the lecture rather than attempting to bypass their schoolwork using the internet. After a few instances of losing points and having a professor email, most students were deterred from answering questions using internet answers.

From the professor's point of view, a flipped classroom gives a lot of flexibility with their class time. Class time can be devoted to answering student questions, doing problems, guided group work, demonstration projects, etc. The downside to a flipped classroom is that the professor has to create a schedule before a semester starts as to what will occur each day and what content needs to be recorded for the outside class video lecture. Creating the video lectures does take time, effort, and planning on the first implementation. After the first implementation though, those recordings can be reused each semester making the flipped classroom preparation cost to the professor much lesser. In a time of pandemic or other emergencies, a flipped classroom does lend itself well to adjusting to the new situations. Students also need access to a computer and internet which may be problematic but as technology becomes more integrated into households, this will become less of a deterrent with time.

Conclusion

A flipped remote classroom was used to create a flexible curriculum during the Covid-19 pandemic. The students showed metacognition and similar academic achievement which may indicate that a flipped-classroom approach may be one of the best ways to adapt the content in extraordinary circumstances. Students seemed to find the approach acceptable once they became used to the differences from a traditional classroom. The downsides to the approach are a large amount of instructor preparation of lecture videos and worksheets along with grading burden. The grading may be alleviated by taking a different approach than this instructor by using online homework software.

References

1. Garcia-Vedrenne, A. E.; Orland, C.; Ballare, K. M.; Shapiro, B.; Wayne, R. K. Ten strategies for a successful transition to remote learning: Lessons learned with a flipped course. *BMC Ecol. Evol.* **2020**, *10*, 12620–12634.
2. Findlay-Thompson, S.; Mombourquette, P. Evaluation of a flipped classroom in an undergraduate business course. *Business Education & Accreditation* **2014**, *6*, 63–71.
3. Mooring, S. R.; Mitchell, C. E.; Burrows, N. L. Evaluation of a flipped, large-enrollment organic chemistry course on student attitude and achievement. *J. Chem. Educ.* **2016**, *93*, 1972–1983.

4. Alley, C. P.; Scherer, S. The enhanced flipped classroom: Increasing academic performance with student-recorded lectures and practice testing in a "flipped" STEM course. *The Journal of Negro Education* **2013**, *82*, 339–347.

5. He, W.; Holton, A. J.; Farkas, G. Impact of partially flipped instruction on immediate and subsequent course performance in a large undergraduate chemistry course. *Computers & Education* **2018**, *125*, 120–131.

6. Andrade, M.; Coutinho, C. Implementing flipped classroom in blended learning environments: A proposal based on the cognitive flexibility theory. *E-learn: World conference on e-learning in corporate, government, healthcare, and higher education, Association for the Advancement of Computing in Education (AACE)* **2016**, *1*, 1115–1125.

7. Simko, T.; Pinar, I.; Pearson, A.; Huang, J.; Mutch, G.; Patwary, A. S.; Ryan, K. Flipped learning–a case study of enhanced student success. *Australasian Journal of Engineering Education* **2019**, *24*, 35–47.

8. Ahmad Uzir, N. A.; Gašević, D.; Matcha, W.; Jovanović, J.; Pardo, A. Analytics of time management strategies in a flipped classroom. *J. Comput. Assist. Learn.* **2020**, *36*, 70–88.

9. Pérez-Sanagustín, M.; Sapunar-Opazo, D.; Pérez-Álvarez, R.; Hilliger, I.; Bey, A.; Maldonado-Mahauad, J.; Baier, J. A MOOC-based flipped experience: Scaffolding SRL strategies improves learners' time management and engagement. *Computer Applications in Engineering Education* **2020**, *29*, 750–768.

10. Peters, E.; Kitsantas, A. The effect of nature of science metacognitive prompts on science students' content and nature of science knowledge, metacognition, and self-regulatory efficacy. *School Science and Mathematics* **2010**, *110*, 382–396.

11. Zohar, A.; Barzilai, S. A review of research on metacognition in science education: Current and future directions. *Studies in Science Education* **2013**, *49*, 121–169.

12. Lee, H. W.; Lim, K. Y.; Grabowski, B. L. Improving self-regulation, learning strategy use, and achievement with metacognitive feedback. *Educational Technology Research and Development* **2010**, *58*, 629–648.

13. Castelli, F. R.; Sarvary, M. A. Why students do not turn on their video cameras during online classes and an equitable and inclusive plan to encourage them to do so. *BMC Ecol. Evol* **2021**, *11*, 3565–3576.

14. Epstein-Shuman, A.; Kushlev, K. . *Lights, Cameras (on), Action! Camera Usage During Zoom Classes Facilitates Engagement Without Increasing Fatigue.* https://osf.io/jq7us/?pid=9uta5 (accessed June, 2021)

15. Oranburg, S. C. *Distance Education in the Time of Coronavirus: Quick and Easy Strategies for Professors*; Duquesne University School of Law Research Paper No. 2020-02; 2020.

Chapter 4

Problem Based Learning Group Projects in an Online Format – A Sequential Approach

Simona Marincean* and Marilee A. Benore

Department of Natural Sciences, University of Michigan—Dearborn, Dearborn, Michigan 48128, United States
*Email: simonam@umich.edu

Student-learning by conducting projects in teams or groups is a well established and successful science practice. In problem or team-based learning students learn and enhance science understanding by application, and they become more proficient in other skills such as collaboration, leadership, communication, problem solving, and brainstorming. Moving group work into an online format is challenging but can be accomplished with consideration of the project goals. When forced to suddenly transition all coursework and projects into online formats we found that the faculty communication and shared interest in student skill development were critical, especially across disciplines where hierarchal program structure means that advanced courses rely on consistent content in foundational courses in supporting disciplines. These conversations influenced strategic development of projects, often on the fly, with specific goals of scaffolding student skills and expectations in creation, redesign and modification of projects to achieve goals. Three courses, two foundational prerequisites required for an upper division course, typical of any chemistry/biochemistry program, are described. This sequential approach, achieved by interdisciplinary faculty coordination of learning objectives and skill development over a series of scaffolded and interdependent courses and content, builds students' group work skills not only in the scientific domain, but also communication and organizational capacity.

Background

Group work has long been a core practice in science and engineering courses. It is driven by the body of evidence supporting group work as a best practice in student engagement and learning (*1*), as a learning objective aligned with and supportive of career training (*2*); and as a practical aspect of laboratory experiments that are complex and require teamwork (*3*). Described as team-based learning (TBL), problem-based learning (PBL) (*4–6*), project-based learning (*7*), (PBL), and course-based research style experience (CURES) (*8*); extensive literature is available to guide faculty.

© 2021 American Chemical Society

The COVID pandemic has resulted in swift conversion of in-class activities into online formats, and the adjustment has likely been the most dramatic in STEM fields. Faculty were required to muster new skills to support teaching and student learning: recording lectures, completely redefining laboratory experiments to incorporate videos (9) or creating new "take-away" at-home labs and kits (10) requiring safe at home materials (11–13). Faculty were encouraged to avoid mimicking the former format into an online version, but rather use the program and course learning objectives as the guides and benchmarks for ensuring content, not delivery methods (14).

Some STEM professional societies reacted quickly, with professional societies such as the American Chemical Society and the American Society for Biochemistry and Molecular Biology (ACS.org, ASBMB.org) at the forefront (15, 16). The STEM societies and their members worked to provide examples, videos, support, and connections. Special issues (17) were published with strategies and examples of moving courses online, including assessment (12, 18–21). Most importantly, faculty reached out to assist each other across disciplines. Facebook groups like *Strategies for teaching chemistry online* were formed to assist and share, offering discipline-based suggestions and support, faculty peer support, and dry humor - all critical to success.

The rapid transition to virtual content required different approaches for lectures, supplemental instruction (SI), office hours, recitations, laboratory overview delivery as well as actual experiments (virtual versus at home) and learning assessment methods. Complicating the efforts were modifications in accepted campus practices brought by the changing rules for safety during the pandemic.

Campuses adjusted to support the lecture infrastructure with online platforms, software and hardware. Moving traditional and successful class and lab PBL into computer-supported collaborative learning (CSCL) (22) resulted in what is likely the greatest number of educational reforms in science. On our campus, the synchronous contact time for a three-credit lecture course like organic chemistry (with pre-COVID mandatory meeting times at three hours/week) was limited to 75 minutes, which was dedicated to more challenging topics and in-class exercises as agreed by the faculty involved in teaching the course. As a result, the learning experience gained a significant asynchronous component. Group work, one of the hallmarks of student-centered learning, was notably negatively impacted since in-person observations and experiences are vital to projects in both laboratory and lecture.

In developing new activities for both lecture and laboratory courses, one of the common goals across disciplines in our Department of Natural Sciences was that students will participate in group projects throughout the curriculum. Team based learning exists in many formats based on length: semester long or short project; group selection: random or by choice; group size; interaction type: live or collaborative documents; review and evaluation: outcome-paper presentation (essay or formal report), oral presentation; assessment; and peer observation. Roberts and McInnerney in their paper (23) identified common challenges regardless of format: student antipathy towards group work; group selection; lack of essential group-work skills; the free-rider group member; student ability inequality; group member withdrawal, and assessment of individuals within groups. We attempted to mitigate the problems with their suggested solutions and incorporate personal teaching experiences into the specific group work structure for the online projects.

Group work is often conducted in laboratory experiments. An obvious and immediate task was how to ensure that the American Chemical Society certification guidelines for "Team Skills", item 7.5 in the certification guide (24) were met. Reproducing laboratory experiences and skills was limited and fraught with problems of transitioning hands-on skills to virtual simulations, or equivalent experiences done at home. Limited access to lab instrumentation, reagents, lack of safety options, and asynchronous delivery challenged both faculty and students, and eliminated the capacity to work in teams at the bench. There were few examples of how to manage successful science group work in online courses, especially laboratory experiences. The skills used in lab activities vary somewhat and are distinct from the ones needed in lecture group projects.

Students were less likely to know each other in online work, especially in introductory classes and were apprehensive about sharing personal information online, although busier as many chose to take more credits or work more hours during the pandemic shutdown. Faculty created kits that were sent home to recreate "kitchen chemistry". Virtual experimental videos were released, experiments were recorded and access temporarily expanded and free at many subscriber resources (*11*). But most of these were not shared experiences.

On our campus we were specifically concerned as our students were very isolated from peer and faculty interaction. Our campus is a commuter PUI (predominantly undergraduate institution) in the Midwest with a high proportion of students who are first-generation college, likely to be Pell supported, and of diverse backgrounds. While the University of Michigan Dearborn is considered one of three campuses of the University, our programs and student demographics are remarkably different from our flagship, the Ann Arbor campus. We have separate budgets and leadership but do have support advantages such as library holdings and software.

Monitoring of student experiences via informal and formal queries were an early catalyst for faculty and administrative reflection. Ensuring successful incorporation of group projects was raised as a concern on our campus. Reports about the lack of faculty-student communication and increased students' isolation were validated by a non-anonymous campus survey conducted in Fall 2020. Over 300 (out of 8500) student surveys were completed, with students documenting problems and obstacles in their coursework. The majority of student respondents to the survey indicated they were white or Caucasian, with about a third students of color. It should be noted that MENA (Middle Eastern/Northern African) statistics were not included, although our campus is set in metro Detroit, the largest concentration of Arab Americans in the US.

Results of the campus survey indicated that during the pandemic nearly half of our students worked, the majority more than 11 hours per week, a third of those more than 20 hours per week, and a third were responsible for caring for family members. The respondent majority indicated appropriate computer and internet access and ease with the online software and learning management system (unlike many faculty). The top benefits students reported about learning remotely included flexibility, lack of commute, and safety. Difficulties included challenges with learning remotely, connection, teaching and communication. In terms of consensus about courses and satisfaction, a slight majority preferred asynchronous recorded lectures. While it is not known how many of our campus courses normally incorporate collaborative work, only 28% indicated their courses required collaborative projects, and a third of those respondents did not like the experience. Students did not indicate that collaborative group work made them feel more connected to the university, as 44% felt very little to no connection with the university, 45% felt some connection, and only 11% felt a strong connection.

Our Approach to Online Group Learning

In this paper we outline the successful transition to online format of three group projects in biology, chemistry, and biochemistry courses at different levels: introductory, gateway, and upper division. This is due in part to the dependence of biochemistry programs on the foundational biology and chemistry courses, so communication is vital in program development and assessment of student learning. This sequential approach of scaffolding group learning skills allows communication and integration of faculty experiences and graduated expectations of students. It also addresses the needs of first-generation students that benefit from more assistance and experience with teamwork, communication, and learning skills compared to second and third generation students who may have family guidance.

In a freshman biology course, a hands-on biochemistry exercise was converted to online team projects, with video and model building presentations; in a sophomore organic chemistry lecture course groups were tasked with development of written summaries and video presentations of relevant topics; and a book and literature review in an advanced biochemistry course was recreated with the face-to-face meeting and poster presentation converted to virtual meetings and video presentations. In each of these courses students had access to either Supplemental Instruction (SI) or capstone student mentoring and support.

The three projects served common education goals which were formulated during the transitions. The students should be able to:

- Present a topic in a logical manner with appropriate amount of detail and relevant examples.
- Search and include necessary information from additional resources outside of lecture.
- Give an oral presentation on a specialty topic.
- Provide peer evaluation on a written draft or presentation.
- Reflect on feedback and incorporate suggestions when deemed valuable.
- Engage in collaboration with a group.
- Read, use and comprehend specialty literature and science databases.

We describe the projects and detail the specific positive outcomes and problems and how these were addressed. Across all projects, many of the challenges were similar to those detailed by Roberts and McInnerney, mentioned earlier in this paper. We have identified several additional ones: new format, unfamiliar to both students and faculty, requiring more oversight than the traditional one; synchronous versus asynchronous content; conversion of hands-on activities to synchronous or asynchronous lab activities (often with less faculty oversight); increased resistance to scheduled synchronous times due to additional workloads or newly arisen family responsibilities due to COVID. The project assessments are presented as a mix of informal formative and summative assessment, incorporating previously used assessment and new tools.

Projects

Introductory Course – Biology

In the introductory biology course students learn the chemistry and biochemistry of cellular molecules and processes. This course is a critical prerequisite for the upper division biochemistry and bioinorganic chemistry courses, amongst others, and typically 60-80 students are enrolled per section. The co- or pre-requisites for the course are the first semester of general chemistry and a specific math competency. It is a four-credit lecture-lab linked course. It covers water and acid base chemistry; amino acids, nucleic acid and carbohydrate chemistry, their polymeric forms (proteins, complex carbohydrates, DNA, RNA) and the processes by which they are produced; catalysis, energy transformation and regulation of processes. Lab topics include hydrophobic and hydrophilic properties, enzyme kinetics, cell biology, fermentation and other processes. Typically, students spend four hours in the lab working in teams of 3-4 students, under the direction of faculty members, (as a PUI we do not have graduate students teaching labs, but we do have supplemental instruction). Students were not allowed in labs for over a year during the pandemic, and the faculty spent the summer modifying and creating kits for students to use at home to simulate as many laboratory experiments as possible. Most labs were asynchronous due to the family and work demands of students. Lectures and labs were virtual, with one of the two lectures per week synchronous. The majority of the lab was attempted to be taught synchronously, with relaxed standards for experimental completion and virtual data "sharing".

The group project was created to assist in understanding of structure/function in macromolecules. Students had difficulty understanding the relationship of chromosomal/genomic information and how it relates to protein structure and function. Students understood the structural concept of the DNA wound into a helix, and created models showing the linkages in labs. They also created protein models linking together amino acids and predicted folding based on hydrophobic, hydrophilic and charge properties of side chains. The PBL project was created to enhance student understanding of the direct relationship of the chromosome structure and DNA sequence and proteins.

A project was developed a number of years ago in which students were assigned in small teams in the lab to create models which were displayed at the end of the term. The idea for the chromosome project stemmed from the Swiss Institute of Bioinformatics 10th anniversary display (25), which was a chromosome "walk" through a park, as well as the magnificent giant chromosome display in Valencia, Spain at Museu de les Ciències Príncipe Felipe (26).

A list of genetic diseases with information about the gene and protein was provided, small groups of 2-4 students were created by the instructor and students selected a topic. Over the course of a few weeks, toward the end of the term, they researched the databases to understand the disease, genetic mutation, gene location on the chromosome, protein function, protein sequence and protein structure (if known). In the online format the groups were provided with opportunities to set up specific Zoom meetings for guidance. The course SI and librarian created simple videos to assist with obtaining information and data from the databases, especially the National Center for Biotechnology Information site of references, genetic information and sequence databases (27).

In non-pandemic times students made a three-dimensional model of the chromosome, and with the accompanying information, created a display of their labelled and properly banded chromosome, indicating the site of the gene with accompanying data, and projects presented in a campus atrium as a showcase event. Individual groups were queried and graded by the faculty member, and students conducted peer evaluations. During the pandemic it became critical to increase group work as these students indicated they felt disconnected. These first- and second-year students in particular needed guidance in teamwork and it was determined to move the project into an online format. They were still required to make models but collaborated virtually by assigning roles within the project to create a virtual display. The presentations and models were incorporated into video presentations and shared to all in the class. Students were required to view several in order to answer essay quiz questions. The stakes were small (a small percentage of the grade) and meant to encourage fun dialogue and to develop teamwork. Students understood that each would be queried to ensure understanding of all parts of the display. Prizes were awarded for most educational, most amusing, most entertaining, most delicious, and most creative. Peer project evaluations (submitted only to instructor) allowed students to provide informal assessment but also provided opportunity to remark on problems or discrepancies in teamwork, often typical in freshman classes where independent and group work skills are less developed.

Results and Outcomes

Students' projects were on par with those on campus, with stimulating displays and creative ventures, such as writing songs lyrics to popular tunes or creating limericks and poems to accompany the chromosome and protein models and presentations. They clearly learned the skills of using data sets and literature to create accurate visual models. In general students appreciated the project, acknowledging it helped their understanding, and reported that they enjoyed working with a group. Students were comfortable working online and although the technical skills were not as advanced as upper-class students, all presentations were professional.

Successes

There were many positive outcomes, and this project could easily be continued in non-virtual labs settings. The change to faculty pre-formed teams, unlike prior years when students worked with lab partners, allowed team members to form cohorts instead of depending on prior groups and friendships, which was difficult during the pandemic. By creating a low stakes option, choosing a gene to be the first group decision, a positive, inclusive setting was generated from the starting point.

Faculty Challenges

The greatest challenge was in ensuring the groups were on task. Intro course students are less likely to be comfortable with their own knowledge, and therefore often hesitant in group settings. The lack of prior group-work skills can set up unfortunate dynamics and student complaints, and are often reported in the peer evaluations, which is often too late to address and change. Finally, the knowledge disparity meant it was critical that faculty carefully monitored groups so that those who lacked conceptual understanding or skills were given an opportunity to ask questions in small groups or individually outside class. This led to a large amount of additional time commitment from faculty and extensive interaction beyond office hours, as well as the creation of additional "how-to" videos on using databases. We were able to use librarian and SI assistance for some of these tasks.

Sophomore Course – Organic Chemistry

The three-credit sophomore organic chemistry course presented is the second part of the one-year sequence geared at students interested in chemistry and health related professions. In a pre-pandemic format, it met weekly during a 14-week semester, twice (75-minute sessions) and once (50 minutes) for lectures and recitations, respectively. The material covered structure and reactivity as a function of functional groups: alkene, alkynes, carbonyl, carboxylic acids and derivatives as well as a brief survey of amino acids, carbohydrates, and lipids. At the COVID outbreak the synchronous meeting times were reduced by 50% to one 75-minute lecture and a noncredit 50-minute recitation. The faculty involved in organic chemistry sequence agreed to deliver the material asynchronously in recorded videos and dedicate the synchronous meeting times to more challenging topics and to problem solving activities, essentially teaching a flipped class. The synchronous meeting times were through Zoom. The weekly student assignments were: quizzes as incentives for the students to watch the recordings before the synchronous lecture; homework as checks for problem solving skills on particular concepts, and group projects as additional study settings and peer discussion of the more challenging topics. While the quizzes and homework were similar to the pre-pandemic course version, the group projects were developed as part of the conversion to online format. The topics for the projects were selected such that it would encourage the use of additional resources besides the recordings of the week. The students were organized in groups of 5-6. At the beginning of the week the assigned group member provided the preliminary version that was then discussed and edited by the whole group with a final version submitted by the end of the week. The author of the preliminary version had to submit an accompanying video presentation. Specific assignment requirements were: length of the summary; max three pages; balanced content of specific chemical examples and discussion; evidence of collaboration from each group member; no evidence of plagiarism; and five-minute-long video presentation. The students interacted outside the lecture or recitation time and it was their responsibility to connect. Grading rubrics were provided for both the summary and video presentation. Upon completion of these projects, the students were able to identify functional groups and infer physical properties and reactivity, write reaction equations and draw their mechanisms with proper electron flow.

Results and Outcomes

The summaries were generally completed on time, with varying degrees of content and detail among the different topics and groups. Generally, the summaries were structured better when the topic aligned with the textbook chapters than when the subject was only a part of the chapter. Despite feedback provided by the faculty and a content appropriateness category in the grading rubric, the students resisted content adjustments when needed, which suggested that they lacked the confidence to eliminate material that was needed and instead opted for extra material, "the more the better". Increased course load, internet access, work overload, and family responsibilities were mentioned as reasons for the situations that led to late submissions. As the semester progressed the summaries included more specific examples, indicative of students' improved understanding of the material and its relevance. At one end of the spectrum, the examples were very similar with the ones from textbook or asynchronous material (more often when the assigned student was not the strongest of the group) and at the other end fully original and detailed. There were few cases when either one student of the group was significantly stronger and chose to complete the assignment before any contributions could be made - suggesting lack of confidence in her/his team partners skills and knowledge, or the student did not engage at all, lacking the self-confidence to make any contribution. However, the majority of the groups worked together, with the assigned student as the project leader for the respective week and each contributing towards an improved version that maintained the original structure and voice. The video presentations allowed the students to showcase the key elements of the topic, identified through group discussion (the five-minute length was set intentionally).

Successes

Students shared verbally that having to organize the material around a specific topic from a chapter afforded an additional setting for studying. Communication about a concept - talking or writing- aids one to check her/his understanding. Being able to identify challenging areas and ask questions or answer another group member's question(s) empowered them to continue. Moreover, the students were able to use the summaries for test preparation as they were assigned before the exam on specific topics. The students communicated on social media outside of the university LMS system setting up WhatsApp chat groups for quick exchanges. That way theory had an informal environment that made them comfortable about participation, providing and receiving feedback. Having to give an oral presentation was an incentive to revisit the material to ensure that there were no mishaps.

Assignments were designed to have the students revisit a subject several times and learn through communication. The students were exposed to each other's method of information accumulation, and dissemination. Students could learn from their peers through feedback, and that trickled in their learning habits, such as time management (they had to start studying early and take good notes on the material - so they were able to participate in preparation of the summary). The students learned that examples used by faculty or in the textbook cannot be considered original and having to provide additional structures or reactions are needed tests of their understanding.

Challenges

Student engagement was not perceived as uniform among group members. The students were asked to highlight their individual contributions in the final paper and thus the faculty was able to evaluate the extent to which they participated in each project.

When probed by the faculty during individual conversations one of the most frequent reasons for lack of involvement was additional commitments: full-time job, family responsibilities exacerbated by COVID-19 pandemic, larger than usual course loads that were time consuming. Students were not able to dedicate enough study time to be able to make meaningful comments or they felt overwhelmed by the material and gave up entirely.

Another contributing factor was lack of confidence. Some students did not trust themselves to risk providing the wrong information or example or they were not able to recover after a harsher feedback. They would need encouragement and they needed it from the group members in order to re-engage. Each group had one or more of the group members that took the lead without giving opportunity to the group members to participate until most of the assignment was completed. Sometimes the "leader" justification was the risk of a low grade, which was actually minimal since these were low stakes assignments. The occurrence of such instances was reported by the students who were concerned that their grade would be affected. Other times, one student would make a suggestion that was not trusted by the other group members due to the student's perceived standing in the class and thus ignored unceremoniously.

The groups tended to assign work as in sections to be divided among group members. Group members occasionally felt that they were given a more challenging area than others and required more time to address it and that generated resentment against the assigner/leader. Last but not least, students felt that coming up with new examples was time consuming and thus when on a time crunch, they relied on instructor provided material although the risks regarding plagiarism were constantly discussed during synchronous meeting times.

In order for these concerns to be addressed, the faculty had to actively engage the students in conversation during synchronous meeting times as a group or individually at a mutually agreed time. It required several initiatives from faculty before a significant degree of trust was established successfully.

Faculty Problems

The encountered problems could be organized in three categories: content, ethics, and communication. The content was not as specific as needed, with the students providing irrelevant information that was in the same textbook chapter or asynchronous material as the assigned topic. After a couple of occurrences, a reminder to that effect was included in the weekly announcement of the assignment. There were instances when content was included from either the textbook or the asynchronous material without proper citation, although the students completed a plagiarism module at the beginning of the course. Conversations around plagiarism were frequent and had to include specific scenarios. There were instances when one group member "borrowed" material from a free source without informing the other members, so they were not aware that content was plagiarized.

Each group had at least one of the following types of participants: micro-managing leader, free rider, and unsure, timid member that needed encouragement. Mindful faculty oversight was needed to ensure that all team members were included in discussions. Significantly more involvement was required from faculty who had to act as a moderator among team members when hurdles were encountered. Examples of such instances include not enough time to work on an assignment because the initial version was shared late in the week; submission of a version that was not completely proofed by team members to avoid lateness penalties; resentment over incorrect material that led to a smaller grade. Due to our university pandemic policy students were able to drop the course throughout the semester and as a result groups had to be reorganized, which required careful planning to maintain a successful dynamic.

Upper Division Course — Nonmajors Biochemistry

The biochemistry project was an adaptation of an in-course group project which had been developed and incorporated over many years. In teams, students read and appraised a nutrition/dietician book. For example, a book on a low carbohydrate approach, or a low fat or low salt diet, from a pre-approved list; they also read the literature on which the book background and recommendations are based and created an evaluative presentation, which traditionally had been an in-class, face-to face presentation and poster session.

The non-majors biochemistry course is a three-credit survey course required of chemistry majors, and an option for other interested students. Prerequisites are Introductory Biology, General Chemistry and Organic Chemistry. The course includes rigorous chemistry kinetics; water, acid and base chemistry; structure/function details of amino acids, carbohydrates, lipids and proteins; and metabolism (proteomics and metabolomics). There is limited coverage of genomics but does include transcription, translation, replication and gene regulation. Most students are juniors or seniors and have some experience with team and group projects.

The course offered in summer 2020 following the COVID outbreak was completely asynchronous and taught in block format over 4.5 weeks with daily lectures recorded and made available in real time or for later viewing, to a team-taught section of 60 students. Interactions among students and with faculty, were either through Zoom or our LMS chat feature. While some of the students already knew each other, many did not or were guest students.

The team project required two parts, designed to integrate macromolecular structure/functions, metabolism and homeostasis via nutritional biochemistry:

- A group project in which teams read and evaluated (support or refute) popular press diet books and created a group presentation in any chosen format.
- Individually written research papers on subtopics related to the book. Topics could be of personal interest but related to the book topic, providing the foundational background for each student to be biochemically prepared to support some aspect of the book.

The project goals included ability to: use literature to support an argument; effectively research using literature and databases; write a research paper; work in a group; apply biochemistry molecules and metabolism understanding (into everyday life); ability to use data and literature to support claims. Groups were required to meet several times and turn in a journal of meeting notes, and each group was required to meet with the faculty member several times. Group guidelines included topics to consider during the reading.

The purpose of the group was to facilitate teamwork and increase comprehension of the material, as well as correlate biochemistry with applied understanding to nutritional and metabolic chemistry. This project has been conducted many times over the past 20 years, but this was the first time that the lectures, group work and final presentation were all presented online.

By requiring students to write a related research paper, with annotated bibliography and draft deadlines spanning the group work, students were forced to be prepared for the group work. Meeting with the faculty member was important in keeping them on track and guiding and validating their insight on the biochemistry of the topic. To facilitate understanding, whenever possible course material included references to nutrition, such as required amino acids, the functional role of minerals and vitamins in enzymatic reaction in metabolism or cell signaling, or a threaded topic such as insulin/diabetes.

A selection of science-based books was made available (as opposed to non-science-based diets, or books which are not supported by literature). Students could create their own groups, with strict group size maintained of 5-6 members. Groups met weekly typically during Zoom or Google Hangout sessions, and often via Snapchat. Students were comfortable meeting online and using technology to create presentations. An advanced student volunteered to assist with literature searches and reading the literature. Special recorded sessions were held on reading the literature and creating an annotated bibliography. Students were assured that literature searches, evaluative analysis and writing, teamwork, poster presentations, and communication skills would not only be valuable, but important for faculty to use in future reference letter writing for professional schools or employment. The project and paper were a significant portion of the grade.

During the group meetings with faculty students were individually asked to report on their section of the book (few actually read the whole book). They were asked to consider in the final presentation: *What is the reason this book was written? What are the qualifications of the authors? Is the diet biochemically sensible? Is there scientific evidence to support the claims? Would you recommend this book? Why? What are the pros and cons of this diet? Is there any risk to this diet?* Finally, references were required for the presentation, often drawn from the papers students used in their research paper.

Assessment of papers and projects were graded with a rubric. The annotated bibliography, due before the optional draft, was graded separately from the paper. Research papers and presentations were individually graded using a rubric. Peer evaluations of group members were submitted to faculty to ensure problems or concerns were documented and to gain project assessment/feedback. In addition, an IRB exempt Qualtrics survey was conducted, with about 1/3 of the class responding. Students were required to evaluate other groups as part of the final exam, ensuring some feedback to each other. (All exams were open book and asynchronous, with a 48-hour window per university requirements.)

Results and Outcomes

Students completed projects on time. The research papers were well written and referenced, although writing formats varied. Topics typically include diseases, drug discovery and mechanism of action, or the intersection of epidemiology and biochemistry. The video projects were well done. All were too long, far exceeding the 10-minute time limit by 100-200%. Length of video was a concern to them, as most students firmly believed more was always better. Most presentations were edited into one file with each student participating in organized Zoom presentations. The large file size meant many were uploaded as YouTube videos.

The Qualtrics survey of the course indicated that most found the book interesting. They learned new information and integrated biochemistry knowledge. About half indicated it helped in making connections to others in class and was valuable in honing professional communication skills. Half the students enjoyed the block format, but those that didn't cited reasons such as the intense nature of the course and excessive time demands. In the survey a number indicated they had not written a science research paper prior to this class, although the majority had read research papers. About half indicated the project improved understanding of knowledge and theory of macromolecules, and importantly was something they could discuss with friends or family. In peer feedback 96% indicated all members made an equivalent effort, and nearly 100% indicated the group functioned well together without issues. The number of self-reported hours spent on the group portion of the project varied wildly, with ranges from 5 hours to 40, and an average of 21 hours total per individual.

Student Successes

Students enjoy this project, and often report back on the value of the research papers many years later. Allowing them to pick teams is valuable as those who are on stricter diets (vegans, vegetarians, diabetics) can band together. The majority appreciated the linked nutritional content to biochemistry, but a few would have preferred to focus on core topics, in part to prepare for professional examinations such as the MCAT.

Fewer complaints were voiced about teams in this project, perhaps because students were more experienced in group work. Most students provided specific comments about the group project, indicating it did indeed integrate knowledge across disciplines, that they loved the group work and it seemed like fun not work, they were proud of their results, it was a unique learning experience, and they enjoyed learning professional skills such as making a poster.

Challenges

A problem was lack of equivalent individual effort, but the majority of groups indicated this was not an issue. A few students dropped the course requiring group reformation, which was distressing for the rest of the team. Although books were recommended, occasionally groups picked a book that turned out to be primarily exercises with insufficient science content, or diets in which they failed to understand the lack of rigorous validating data. A rare issue was plagiarism by one member of a team, and it was handled gently by the faculty member to avoid embarrassment. While students should know to not plagiarize, in group work it is best to address it as an example of how to write properly (and stay out of trouble) than be accusatory, a dynamic which can undermine group teamwork.

Conclusions

Prompted by COVID pandemic, group project activities were developed for an online format in courses throughout the chemistry and biochemistry curricula. The students enjoyed group working and shared responsibility but did not feel more connected to the university community as a whole, but rather to their group. The assignments were additional opportunities to discuss and study material in an interactive and less formal environment with the ultimate goal of improving performance, writing and communication skills. However, there was some indication that students were creating cohorts that would be valuable later. *The incorporation of sequential PBL was positive but will require long term assessment to verify its value.* Clearly teamwork is a fundamental skill for a successful career in science and engineering, and is taught through group work activities in courses, lecture and laboratory.

Faculty involvement significantly increased outside of the official meeting times in order to ensure progression of the assignments and mediate instances related to communication, ethics, and mechanics of group functioning. In the future, inclusion of such components could contribute to an involved experience. Perhaps both formal and informal conversations and mentoring would be beneficial to both students and faculty development. The simplest mechanism to that effect would be topics that are incorporated in labs when possible, and these are typically consistent regardless of the course instructor. However, in this specific example, the projects were part of the lecture topics in this approach, since during the pandemic laboratory experiments were disrupted and altered in content more than the lecture topics. If the projects were incorporated as part of a linked lab or recitation, faculty, graduate students or teaching assistants could assist with development and assessment.

This sequential approach to creating group work, in which faculty communication was increased by motivation and common concern about students' skills, was invaluable in considering the overall curriculum experienced by students. It should be noted that this collaboration is more critical in STEM courses, as many liberal arts programs have less hierarchical structural design. It is important for faculty to understand the learning objectives of the courses within a program. A course is often the prerequisite for courses within other majors, and therefore cross-discipline discussions, while often difficult to manage, are important in overall program assessment. The student learning experience should seem seamless and supportive in their program. By considering the overall learning objectives, not just science content in creating programs, the full student experience would be integrated across the program span.

In summary, a sequential approach in converting the PBL projects demonstrated the utility of the online approach, increased faculty collaboration and resource sharing; all critical in developing faculty teachings skills and advancing pedagogical awareness. It aids in facilitating conversations about the foundational course content, with the ultimate goal of students' preparation for both advanced course content and teamwork skill practice.

Notes

Partial data was obtained under IRB HUM00182941, exempt status.

Acknowledgments

We thank the students in our courses who have participated in the projects with enthusiasm, and who have provided valuable feedback and suggestions. We thank Dr. Rachel Pricer for insightful conversations.

References

1. Johnson, K. A. D.; Smith, S. W.; Sheppard, D. T.; Johnson, R. Pedagogies of Engagement: Classroom-Based Practices. *The Research Journal of Engineering Education* **2005**, *94*, 87–97.
2. Robles, M. M. Executive Perceptions of the Top 10 Soft Skills Needed in Today's Workplace. *Business Communication Quarterly* **2012**, *75*, 453–465; DOI: 10.1177/1080569912460400.
3. Šerić, M.; Praničević, D. G. Managing Group Work in the Classroom: An International Study on Perceived Benefits and Risks Based on Students' Cultural Background and Gender. *Management: Journal of Contemporary Management Issues* **2018**, *23*, 139–156; DOI: 10.30924/mjcmi/2018.23.1.139.
4. Hmelo-Silver, C. E. Problem-Based Learning: What and How Do Students Learn? *Educational Psychology Review* **2004**, *16*, 235–266; DOI: 10.1023/B:EDPR.0000034022.16470.f3.

5. Barrows, H. S. A Taxonomy of Problem-Based Learning Methods. *Medical Education* **1986**, *20*, 481–486; DOI: 10.1111/j.1365-2923.1986.tb01386.

6. Lewis, S. E.; Lewis, J. E. Departing from Lectures: An Evaluation of a Peer-Led Guided Inquiry Alternative. *J. Chem. Educ.* **2005**, *82*, 135–139; DOI: 10.1021/ed082p135.

7. Eberlein, T.; Kampmeier, J.; Minderhout, V.; Moog, R. S.; Platt, T.; Varma-Nelson, P.; White, H. B. Pedagogies of Engagement in Science: A Comparison of PBL, POGIL, and PLTL. *Biochem. Mol. Biol. Educ.* **2008**, *36*, 262–273; DOI: 10.1002/bmb.20204.

8. *Classroom Undergraduate Research Experience (CURE) survey*. Grinnell College. https://www.grinnell.edu/academics/resources/ctla/assessment/cure-survey (accessed 2021-02-27).

9. Marincean, S.; Scribner, S. L. Remote Organic Chemistry Laboratories at University of Michigan-Dearborn. *J. Chem. Educ.* **2020**, *97*, 3074–3078; DOI: 10.1021/acs.jchemed.0c00812.

10. Schultz, M.; Callahan, D. L.; Miltiadous, A. Development and Use of Kitchen Chemistry Home Practical Activities during Unanticipated Campus Closures. *J. Chem. Educ.* **2020**, *97*, 2678–2684; DOI: 10.1021/acs.jchemed.0c00620.

11. Buchberger, A. R.; Evans, T.; Doolittle, P. Analytical Chemistry Online? Lessons Learned from Transitioning a Project Lab Online Due to Covid-19. *J. Chem. Educ.* **2020**, *97*, 2976–2980; DOI: 10.1021/acs.jchemed.0c00799.

12. Fergus, S.; Botha, M.; Scott, M. Insights Gained during COVID-19: Refocusing Laboratory Assessments Online. *J. Chem. Educ.* **2020**, *97*, 3106–3109; DOI: 10.1021/acs.jchemed.0c00568.

13. Baker, A. J.; Dannatt, J. E. Maintaining an Active Organic Class during the COVID-Induced Online Transition at Two Undergraduate Institutions. *J. Chem. Educ.* **2020**, *97*, 3235–3239; DOI: 10.1021/acs.jchemed.0c00759.

14. Procko, K.; Bell, J. K.; Benore, M. A.; Booth, R. E.; del Gaizo Moore, V.; Dries, D. R.; Martin, D. J.; Mertz, P. S.; Offerdahl, E. G.; Payne, M. A.; Vega, Q. C.; Provost, J. J. Moving Biochemistry and Molecular Biology Courses Online in Times of Disruption: Recommended Practices and Resources - a Collaboration with the Faculty Community and ASBMB. *Biochem. Mol. Biol. Educ.* **2020**, *48*, 421–427; DOI: 10.1002/bmb.21354.

15. *Online teaching: Practices and resources*. American Society for Biochemistry and Molecular Biology. https://www.asbmb.org/education/online-teaching (accessed 2021-02-27).

16. *Teaching Labs in the Times of COVID-19*. American Chemical Society. https://www.acs.org/content/acs/en/education/policies/acs-approval-program/events-resources/teaching-labs-covid.html (accessed 2021-02-27).

17. Special issue: Insights Gained While Teaching Organic Chemistry in the Time of COVID-19. *J. Chem. Educ.* **2020**, *97*, 2375–3470.

18. Fontana, M. T. Gamification of ChemDraw during the COVID-19 Pandemic: Investigating How a Serious, Educational-Game Tournament (Molecule Madness) Impacts Student Wellness and Organic Chemistry Skills While Distance Learning. *J. Chem. Educ.* **2020**, *97*, 3358–3368; DOI: 10.1021/acs.jchemed.0c00722.

19. Soares, R.; de Mello, M. C. S.; da Silva, C. M.; MacHado, W.; Arbilla, G. Online Chemistry Education Challenges for Rio de Janeiro Students during the Covid-19 Pandemic. *J. Chem. Educ.* **2020**, *97*, 3396–3399; DOI: 10.1021/acs.jchemed.0c00775.

20. Tan, H. R.; Chng, W. H.; Chonardo, C.; Ng, M. T. T.; Fung, F. M. How Chemists Achieve Active Learning Online during the COVID-19 Pandemic: Using the Community of Inquiry (CoI) Framework to Support Remote Teaching. *J. Chem. Educ.* **2020**, *97*, 2512–2518; DOI: 10.1021/acs.jchemed.0c00541.
21. Nguyen, J. G.; Keuseman, K. J.; Humston, J. J. Minimize Online Cheating for Online Assessments during Covid-19 Pandemic. *J. Chem. Educ.* **2020**, *97*, 3429–3435; DOI: 10.1021/acs.jchemed.0c00790.
22. Dillenbourg, P.; Fischer, F. Basics of Computer-Supported Collaborative Learning. *Zeitschrift für Berufs- und Wirtschaftspädagogik* **2007**, *21*, 111–130.
23. Roberts, T. S.; McInnerney, J. M. Seven Problems of Online Group Learning (and Their Solutions). *Educational Technology and Society* **2007**, *10*, 257–268.
24. *ACS Approval Program.* American Chemical Society. https://www.acs.org/content/acs/en/education/policies/acs-approval-program.html (accessed 2021-02-27).
25. *Chromosome Project.* Swiss Institute of Bioinformatics. https://www.sib.swiss/what-is-bioinformatics/past-outreach-activities (accessed 2021-02-27).
26. *Museu De Les Ciences.* https://www.cac.es/es/museu-de-les-ciencies/museu-de-les-ciencies/descubre-el-museu.html (accessed 2021-02-27).
27. *National Center for Biotechnology Information.* https://www.ncbi.nlm.nih.gov (accessed 2021-02-27).

Chapter 5

Maintaining Rigor in Online Chemistry Courses - Lessons Learned

Mitzy Erdmann,[1,*] Sithira Ratnayaka,[2] Brenna A. Tucker,[1] and Elizabeth Pearsall[3]

[1]Department of Chemistry, University of Alabama at Birmingham, Birmingham, Alabama 35294-1240, United States

[2]Department of Chemistry, Oglethorpe University, Brookhaven, Georgia 30319, United States

[3]Institute for Teaching Excellence, York Technical College, Rock Hill, South Carolina 29730, United States

*Email: merdmann@uab.edu

Maintaining rigor in an online chemistry course is of the utmost importance to ensure students acquire equivalent knowledge when compared to an in-person, on-ground course. Multiple approaches can be employed to ensure online chemistry courses maintain rigor and academic integrity in the online chemistry classroom. Best practices regarding the structure of remote and online courses are discussed. Types of questions utilized in online assessments, assessment settings that discourage cheating, and the use of online proctoring will be shared. A variety of methodologies utilized, course structure, and strategies for maintaining rigor and academic integrity in introductory and general chemistry courses at multiple institutions will be presented.

Introduction

The abrupt move to remote instruction that students and faculty had to endure in Spring 2020 can best be described as chaotic, leaving students upended and faculty scrambling to learn new ways to reach students (1, 2). As the shift to remote learning as an emergency measure was never meant to mimic online learning, both content delivery and the effectiveness of it varied widely (3, 4). Providing students with rigor, both in course presentation and in assessment, was one means of lessening the chaotic feel of the quarantine and increasing the value of the learning experience. Carefully planned course design can also "force" students into improving their time management and organizational skills, as reported by faculty at Mercer University (5). This paper aims to describe four authors' thoughts on what it means to offer "rigor" to their students and the steps they took to instill it in their courses and institutions.

Though many will immediately shift their thoughts to assessments when they hear the term "rigor", delivery of course information is arguably the instructor's first line of defense in terms in this area. If remote and online students are expected to perform equivalently to on-campus students, it should also be expected that the content delivery is equivalent across all modes. While online courses continue to grow in popularity and allow institutions to potentially reach more students not all students have the self-discipline to succeed in such a self-paced environment (*6–10*). Adding the perceived difficulty of chemistry to that creates a less-than desirable situation for students of the subject. This can be circumvented by meeting synchronously with students and student interaction can be increased by adding a response system to these meetings (*11*).

Though we feel rigor begins with course design, it is important to assess students and to make that assessment meaningful. Not only is the content of the assessment critical, as it drives student perception about what is important in a course, but the delivery of the assessment is also important. Though there are a seemingly endless number of ways to assess students, the goal of any assessment is to aid the instructor in extracting what students know and can do (*12, 13*). The evidence gained from these assessments is more robust when they are structured to mimic the task students need to complete (*14*).

A time-tested form of assessment is the exam, which is common among all delivery platforms and varies widely across institutions. Online testing platforms such as learning management systems and online homework systems facilitate the use of multiple-choice exams in addition to fill-in-the-blank, short answer, and video exams. It has been reported that instructors rely more heavily on multiple choice questions than free response in an online setting (*11*). Lower stakes assessments, such as pre-class quizzes, can be developed to support both synchronous and asynchronous learners during interruptions in instruction (*15*).

With an increase in the number of online and remote courses comes reports of increased cheating on quizzes and exams (*16*). ProctorU, a company that provides live proctoring services, cites that instances of cheating have increased from roughly 1% pre-pandemic to greater than 8% post-pandemic. Thus, a portion of students have taken advantage of the increased number of online exams and cheated in a number of ways, as simple as using search engines to look up correct answers to as brazen as paying "ringers" to take the exam for them. While no online monitoring method is going to 100% eradicate cheating - the same can be said for in person methods - the instances of cheating can be minimized using a variety of testing settings and techniques. These include requiring students to agree to honors code statements, using lockdown browsers with or without recorded monitoring, and live in-person monitoring. Use of these services can help to minimize cheating and level the playing field for those students who are taking the moral high road and avoiding cheating (*17*).

As discussed, there are a variety of ways to introduce rigor to a course and a number of considerations before designing. Three institutions: The University of Alabama at Birmingham (mid-sized, public, four-year), York Technical College (public, two-year), and Oglethorpe University (private, four-year) have worked to share their approaches to rigor in course design and assessment. We discuss how courses were initially moved to remote in Spring 2020 and what changes were necessary to improve student learning and assessment.

Institution Backgrounds

UAB Introduction

The University of Alabama at Birmingham (UAB) is a public, mid-sized university associated with a major R01 research institution. Undergraduate enrollment at UAB has increased over the past five years with a record enrollment of 22,563 students enrolling in the Fall 2020 semester. The "undergraduate-side" of campus is affiliated with a major, multi-hospital medical center and attracts a large number of students who aspire to join the medical field. Because of this large pre-health population and the prerequisite nature of introductory and general chemistry courses, the Chemistry Department at UAB services a sizeable portion of the student population, with an average of 550 students enrolling in the Introductory (non-major) Chemistry sequence (CH 105 and 107) and 1000 students enrolling in the General (science-major) Chemistry sequence (CH 115 and 117) each term. Enrollment in these courses has grown steadily over the past five years and during the spring 2020 semester there were 589 students enrolled in the Introductory Chemistry Sequence and 1,059 students enrolled in the General Chemistry Sequence.

UAB's spring 2020 break was scheduled to begin on Monday, March 16, 2020. The University and the University of Alabama System had issued travel restrictions as early as March 4 and by March 12 had distributed an academic continuity plan that included cancelling in-person classes and meetings. But due to the uncertainty surrounding the novel coronavirus at that time, students were dismissed from campus on Friday, March 13 with every indication that they would return before the end of the spring term. However, the state of Alabama saw a spike in infections following the announcement of its first confirmed cases of the novel COVID-19 virus on March 13 and the University quickly shifted gears, moving to limited business operations and restricting access to campus wherever possible. These restrictions included both faculty and students and meant that a number of students who had left belongings on campus, including course materials, no longer had access to them.

In addition to the difficulties that a lack of course materials posed to both students and faculty, the disparity among the student population made content delivery in a pandemic problematic. In the fall of 2019, only 16.7% of students at UAB claimed a permanent residence outside of the state of Alabama. While 42.1% of students reside in the counties that house or surround UAB, which are modernized and generally offer technological amenities, the remaining 41.2% of students come from other Alabama counties. Many of these counties are rural, sparsely populated, and lack sufficient access to public and/or private internet connections. While local internet providers offered low or no-cost internet access for the remainder of the academic year, these offers did not always reach the most rural of students. Those that were eligible often had to wait weeks for service to be connected.

Oglethorpe Introduction

Oglethorpe University is a small sized liberal arts college in Georgia. The majority of first-year students live on campus (84%), and the most populous majors are the basic sciences (physics, biology). The usefulness of maintaining academic rigor regarding course materials and student accountability via online courses is evident when presented within a smaller community, such as that of Oglethorpe University's 1400 students. The faculty to student ratio is 15:1, and introductory chemistry courses are capped at 25 students which affords more personal interactions between instructors and students when compared to larger classrooms. The small classroom setting lends itself towards increasing accountability and maintaining academic integrity. As the majority of instructors are able to build rapport with their students (given the smaller overall number of students per class), they are able to personally check on submitted work for fallacies in academic honesty (a luxury which is not allowed for institutions of larger scale). This is useful for the large amount of first-generation college students, especially since 94% of all incoming students are beneficiaries of financial aid.

York Technical College

York Technical College is one of sixteen colleges in the South Carolina Technical College System serving approximately 4,500 full and part-time students in Chester, Lancaster, and York Counties. The majority of students (70%) are enrolled in career and technical career programs with the remaining students focused on university transfer courses. All students are commuter students and most have external work and family obligations. Additionally, many students are first-generation college students and over 60% qualify for Pell grants. Three chemistry classes are offered at York Tech: General Chemistry I and II, and a one semester General, Organic, and Biochemistry course, all with labs. There are approximately 300 students enrolling in these chemistry courses annually and the majority of these students aspire to enter healthcare programs.

In mid-2019, the College recognized a need to establish consistent expectations for online course layout, design, and teaching expectations. A committee of faculty led by the Institute for Teaching Excellence was convened to develop consistent expectations for online courses across the college. In February 2020, the Quality Learning Council and academic leadership approved the proposed expectations from the committee. Plans were developed to implement these expectations and start the college-wide transition to consistent expectations during the Summer 2020 term. When the COVID-19 pandemic struck in mid-March, these expectations were leveraged to provide all faculty with best practices for online courses to assist in the required emergency remote online transition. To further support faculty, the Institute for Teaching excellence developed training for all faculty around best practices in online courses including methods to create a welcoming and inclusive online classroom, tips for maintaining rigor, and remote proctoring among other topics.

On March 12, 2020, York Technical College decided to move all courses to the online environment temporarily. However, on March 15, 2020, an executive order was issued by the South Carolina Governor requiring the closure and move to remote education for all public K-12 schools and colleges until further notice. At that time, the college cancelled instruction for the week after Spring Break (week of March 16, 2020) to provide 10 days for faculty to transition all courses to the D2L learning management system.

Rigor in Course Design

UAB Gen Chem Course Structure, Spring 2020

Though online course offerings at UAB are numerous and even include the Introductory Chemistry sequence, there have never been online offerings for the General Chemistry lecture, recitation, or laboratory courses. Traditionally general chemistry is taught in large classes of 185-275 students with instruction delivered via Socratic lectures. Co-registration in a recitation section, where students complete extra-practice in a small group setting facilitated by peer teaching assistants, is required for all students each term. As was the case with many faculty nationwide, lack of existing online course materials left our instructors scrambling to build and deliver content and online-only assessments with only a few days preparation, creating at roughly the same pace that students were learning for both the lecture and the recitation.

Lecture delivery method varied by general-chemistry instructor, largely based on their confidence level in creating online content and manipulating Canvas, the University's learning management system (LMS). All three delivery modes that follow included posting the recordings of the Zoom sessions to the respective Canvas page. One instructor changed little in their delivery, sticking with their existing PowerPoints and delivering a very traditional lecture via Zoom. A second instructor also started with this approach but provided their students content videos that were created by the author. The author chose to flip their classroom, creating instructional videos students would watch prior to attending class and utilizing class time for extra practice and reiteration of key topics. In this case, "flipped" will refer to students preparing outside of class by watching recordings of the slides presented in a traditional lecture before the scheduled class and working on extra practice problems during the synchronous class meetings. The latter approach will be the focus of the rest of this text.

The decision to flip both the general chemistry I and II classes was based chiefly on the lack of consistent, reliable internet for a sizeable portion of the enrolled students and the fact that many students now had unexpected distractions. Many had been suddenly thrust into the role of caretaker, ensuring that younger siblings completed K-12 schoolwork or that older parents and/or grandparents were cared for. Many had to buy upgraded tech to successfully complete the term, which also meant they had to work extra hours to pay for it. These work hours often overlapped with classes. And even those that had little outside responsibility were sharing wireless networks with working parents who had also been upended by the pandemic. Presenting the material in a guided but largely self-paced manner was the most equitable option the author could devise.

Existing PowerPoint presentations were modified to better fit a delivery without an instructor present. This included increased animation, explained images/graphs, additional practice problems, highlighting of key points, etc. Scripts were prepared in advance of recording to make the most professional product possible in the constrained timeline. The screen-recording option of an iPad was used to create voice-over PowerPoint videos. This allowed the instructor to use the prepared PowerPoint slides and scripts as well as annotate the slides using an Apple Pencil in real time. Videos were posted to Modules in the LMS no later than 24 hours prior to the scheduled class time. Students were instructed to view these videos and take notes prior to the class meetings. Students appreciated the structure this format provided and complied almost immediately. Within the first week the vast majority of the students had adjusted their schedule and were preparing as directed.

Synchronous class meetings were held during the regularly scheduled time via Zoom. Up to 300 students could attend a single Zoom meeting, which was larger than any individual scheduled lecture, but cross-attendance between the different sections was minimal as the Zoom meetings were posted directly to Zoom link embedded in the LMS, were password protected, and required an authenticated (associated with an UAB email) account. These meetings were instructor-led extra practice sessions. Students had prior access to the problems, as they were posted to the learning management system prior to the synchronous meeting. Students were instructed to attempt these problems prior to attending class.

The instructor was an attendant in the meeting twice, once as host using UAB credentials and once as a guest using a personal Zoom account. The meeting was hosted from a dual-monitor PC with external webcam and mic. The PowerPoint slides were run from this PC and the chat dialogue and participant windows were kept open in the unshared window so that questions could be answered in real time. As it was difficult for the instructor to manage all of the moving pieces of Zoom and PowerPoint while offering instruction, teaching assistants also attended these lectures to help monitor the chat and to speak up when the instructor overlooked something. The iPad was utilized so that slides could be annotated and problems worked in real time, which better paced the author than animated slides. The second account was made a co-host so that there were no issues with screen-sharing or annotation permissions. As security and Zoom bombing was an issue, the security settings shared earlier were put into place. Aside from those, the settings were mostly the standard Zoom settings.

All Zoom sessions were recorded to the cloud and were sent directly to Kaltura, the media manager application used by our LMS. There was a great deal of lag time between the synchronous meeting and the availability of the recordings in Canvas, but once they were available the recordings were posted to Canvas. This allowed students who had other obligations to watch them at their leisure and for those that did attend to re-watch them if necessary.

The recitation sections also met via Zoom post-pandemic. Here, groups of 24 students were given initial instructions by a peer teaching assistant. They were then placed in random breakout rooms and asked to complete extra practice problems in small groups. This moved the model that was in use before the pandemic online as seamlessly as possible. Students earned points toward their lecture grade for the work completed during these recitation sections, and as such attendance during the pandemic was equivalent to a standard, in-person term.

Oglethorpe Course Structure

At Oglethorpe, the COVID-19 response was to go fully remote immediately, and faculty were given many "work from home" electronic tools. These included drawing pads, laptops, docking stations, etc. The online introductory chemistry courses were held using a lecture/question/answer format. For example, professors would lecture on a topic, present a worked example to the students, and then allow time for students to work on a practice problem in small groups. During the group work sections, students were moved from the main online session into smaller private sessions so that they could discuss and work on chemistry problems with their classmates, providing feedback and allowing for understanding via practice. After allowing students to work in their groups for a set amount of time, they were called back to the main online session and the answers would be discussed. To motivate all students within the group to work towards solving their practice problem, groups would receive different questions and a member from each group was chosen at random to present their solution to the class. In addition, students could ask for help from the instructor/teaching assistant via a virtual "raise hand" function. This method of lecture held students accountable for learning and understanding the content on an individual basis. Students would not know the identity of the group member that would be called upon to present their solution and thus all members of the group were eager to work on solving the practice problems. This method allows for student engagement and responsibility during class time and discourages students from idly and passively watching class. These methods appeared to contribute greatly to student engagement and success, as shown by the end-of-year course evaluations; students specifically indicated that the problem sessions kept their attention. The ability to implement this type of rigor in course design is made possible by the small class sizes at Oglethorpe. In this setting, instructors are able call on individual students, and by the end of the semester, each student has been given many opportunities to display their understanding of the material.

York Technical College

At York Technical College, chemistry courses are four semester credit hours and are instructed as a paired lecture/laboratory course. Prior to March 2020, all chemistry courses were taught in an instructor - led format face-to-face with a cap of 24 students. Across course sections, consistency is maintained through standard course learning outcomes, point breakdowns, percentages of assessment types, and ensuring each section contains the same laboratory experiments and number of tests. Each chemistry course is broken down into four content modules, with a module test and cumulative final exam required; however, faculty are encouraged to utilize low-stakes formative assessments and periodic quizzes throughout the course.

During the emergency remote transition to online instruction in the Spring 2020 term, faculty were expected to conduct all scheduled class meetings virtually through the Bongo Virtual Classroom embedded into the D2L learning management system. During these sessions, students were afforded the same lecture content they would have been exposed to in an in-person course. Graded assignments, homework, quizzes, and tests were administered through the D2L online classroom environment, ensuring that consistent expectations and rigor were maintained during the time of emergency remote instruction. Faculty were also expected to maintain their standard office hours – 8 hours per week – via Virtual Classroom. Due to limitations with the embedded Virtual Classroom environment in D2L, the college moved to Zoom during the Summer 2020 academic term.

Rigor in Assessment

UAB Academic Rigor, Spring 2020

UAB holds contracts with a number of proctoring services, from simple lockdown browsers to recorded proctoring and live proctoring. However, spring 2020 faculty had been instructed that, whenever possible, outside proctoring services were not to be used. The Provost felt that the sudden shift to remote learning was jarring enough for students and that the addition of proctored online exams was more than the majority of students had signed up for that term. With this in mind, the general chemistry faculty made the decision to offer exams through Canvas without proctoring. Instead, the time given to students was quite tight, if not too short, so that there would not be sufficient time to cheat. An Honor Code statement was included as question 1 on the exam and students were instructed that if their grades for exams 3 and 4 were considerably higher than their average on the first two exams that they would have to meet with the instructor. Averages were adjusted, if necessary, once all students had completed the exam.

Four hourly exams and the cumulative ACS (American Chemical Society) final exam had been planned at the start of the spring semester. Courses were changed to remote instruction before exam 3 and 4, and those were given on their regularly scheduled dates. Students were given between 8am and 5pm CST on those days to complete their exams. The wide window was to ensure that students in various time zones had ample time to complete their exam during "normal" hours. Students who were in other countries that even the 8am-5pm window did not accommodate were assigned a unique exam time based on their needs. In an effort to keep the testing environment as close to pre-pandemic conditions as possible, students were given 50 minutes to complete the exam, the same amount of time they would have had in-class. At the request of the registrar, students who did not originally register for an online or remote course were not forced to use online proctoring services.

The majority of the exam questions were multiple choice, as they were with exams I and II, though various other types of question formats that Canvas offers were used where appropriate. Exam banks were used with multiple questions for each topic to increase variety among students and limit cross-talk/cheating. Where appropriate, instructor-created images (usually hand drawn on the iPad to ensure exclusivity) were used as the basis for questions. Algorithmic questions were used whenever possible, as 200 individual number sets can be created in Canvas with the creation of a single question. Questions often utilized generic reactions (A + 2B $\rightarrow$ C, etc.) or fictional elements (E, M^{2+}, etc) to make them harder to find through an internet search.

Because of the secure nature of the ACS final exams, we could not offer a normed version of an ACS to our students. Instead, a cumulative, non-ACS final exam focusing only on material specific to the course in question was created. The format and delivery of the final was the same as that for Exams 3 and 4. Students could opt to use the average of their first four exams in place of the final exam score if they felt that grade was sufficient.

Oglethorpe Academic Rigor During Assessment

Assessment during online classes presents another unique challenge, especially with regards to ensuring that questions are both fair and thought provoking, while remaining unique enough to be unable to be answered via simple internet search. Indeed, one of the biggest concerns among faculty has proven to be the difficulty of proctoring exams and other assessments, along with procuring viable questions for the modality in which we deliver the exams (online). Personally, questions have leaned towards more pictorial in nature compared to in-person exams, relying on more understanding of custom-made images and figures; this greatly reduces the possibility that students can easily confer with online sources to obtain the answer. While this presents an increase in effort on the exam-writer's part, it has proven to be effective. End of the year exams presented last academic year (in May of 2020) showed a highly skewed grade (40% of all students received an "A", with N>100). Comparatively, the final grades for the Fall 2020 online semester (after thorough planning for a completely online semester) followed a normal distribution (~10% receiving an "A" with N>100). Assessments were also written with strict timing guidelines, which significantly reduced excess time which would allow for extraneous internet assistance. Students taking online exams after the spring 2020 semester were given 1.5 hours to complete the exam, which was the same amount of time as the previous year's in-person class. Questions which utilized hypotheticals also assisted in ensuring that academic honesty and rigor was maintained. For example, theoretical molecules were presented for atomic orbital theory; this way students were not able to simply utilize online resources to procure answers and had to withhold previous semesters' academic standards regarding answers.

While these steps were taken to ensure that online examinations still upheld the rigor of assessment expected of students during the normal in-person semester, there were some steps taken to ensure equity regarding the online assessment and proctoring system. Students who had technological issues, or lacked the proper technology (laptops) were given laptops by Oglethorpe University for use for school (these were the same ones which were given to faculty and staff). Also, students who expressed reluctance on doing classes at home due to issues (usually internet issues) were allowed to stay on campus during the Spring 2020 and Summer 2020 semesters. For the Fall 2020 semester, students were allowed to use a VPN (if desired) to allow for better connections. In practice, the vast majority of general chemistry students did not have issues with connectivity in the Fall 2020 semester.

<u>Utilization and effectiveness of proctoring:</u> While proctoring was utilized via camera-based zoom video meetings, the usefulness of such methods remains to be seen. Possibly this monitoring system was most helpful for students needing to ask questions of the instructor during examinations. No additional screen-monitoring software was utilized during exams, due to financial restrictions presented (along with concerns regarding effectiveness and usefulness of administration).

York Technical College

Prior to COVID-19, York Technical College did not have any chemistry course developed for the online environment. Additionally, all quizzes, tests, and final exams were administered in person prior to the Spring 2020 term. In the general, organic, and biochemistry course, instructors are encouraged to develop their own questions or adapt questions from the publisher test bank, instead of simply using pre-developed publisher test bank questions. In the general chemistry I and

II courses, a peer-reviewed open education textbook (OER) is used and the provided instructor resources do not include a test bank. In these courses, instructors design and develop their own test questions. Across all three chemistry courses, instructors are continuously encouraged to design content and level appropriate tests which employ a variety of question types including multiple-choice, short answer, and problem solving.

Although courses were not developed for the online environment, face-to-face instructors may elect to administer tests online using remote proctoring. Once all instruction was moved to the remote online environment in Spring 2020, instructors were advised to use Honorlock remote proctoring for any assessment that would typically have been administered as a closed-note or closed-book quiz or test in the on-ground classroom. At the start of each Honorlock enabled assessment, students were required to show picture identification and conduct a 360° room scan to demonstrate the environment in which they would be taking their quiz or test was free of inappropriate information and materials. Many remote proctoring settings were automatically enabled including disabling printing and copy/paste functions, webcam and screen recording, monitoring for other computer applications, and the lockdown feature ensuring additional browser windows were not opened. However, instructors were able to disable specific settings if they deemed it appropriate for the assessment, and also maintained the ability to whitelist specific URLs, allowing students to access approved websites. Students were provided with information regarding remote proctoring sharing details of the overall expectation that the remote proctoring environment should match an in-person testing environment to the fullest extent possible. Instructors were also provided guidance that they should consider using appropriate time limits, developing a large test pool in which to pull questions for a given assessment, and that only displaying one question per page could help prevent cheating for quizzes and tests taken in online courses.

Rigor in Lab

UAB

Prior to campus closing in March 2020, graduate teaching assistants in the General Chemistry sequence recorded themselves completing the laboratory experiments that remained on the syllabus. These recordings were high enough quality that students could make accurate measurements on glassware and balances, which they were instructed to complete in their notebook prior to the regularly scheduled lab meetings. Students had been required to complete pre-laboratory notebooks and turn them into their TAs to be scored prior to the pandemic, so this was not a new task for students. This submission requirement remained in place after the campus was emptied and was simply moved online. Lab sections met with their TAs via Zoom, where they received short lectures from the TAs before entering breakout rooms with their groups. Groups completed calculations based on their recorded measurements. They also prepared very simple group lab reports.

In contrast to the remote lab offerings for the General Chemistry lab sequence, students enrolled in the Introductory Chemistry lab sequence (CH 106 and 108) completed hands-on experiments using a commercially available lab kit starting in the Fall 2020 semester. Due to increasing enrollment within the upper-level lab courses of the chemistry department, it was decided to convert the Introductory Chemistry lab to an online version to make more space available in the teaching labs for chemistry majors. The switch to using lab kits in the Introductory Chemistry sequence was an incremental process with course development beginning in the Fall of 2018. During the spring 2020 semester when the COVID pandemic hit, there was a single section of each CH106 and 108 labs that were making use of the lab kits (80 students out of 521 total). University guidelines required all summer 2020 courses to be either fully online or remote, and as such, all sections of the Introductory Chemistry lab sequence were offered online from that term on.

This full transition away from in-person labs as mandated by the UAB just so happened to coincide with the original timeline for a full transition to online labs. In theory, students should not have had problems with the switch to online labs because the online transition of these labs was planned pre-pandemic. In actuality, the pandemic led to a major disruption in the distribution of the lab kits. This was due to an inability to source materials for the lab kit that were more needed for healthcare workers (i.e., gloves). As of the Spring 2021 semester, the disruption in the supply chain has seemed to resolve itself and the distribution of lab kits has returned to its pre-pandemic efficiency.

In the online lab, lab experiments are broken into Modules and students are presented with a safety overview, technique videos, and TA-created content videos that are specific to each lab. Safety is a huge component of the online labs and students were required to pass a safety quiz for each wet lab before gaining access to the procedures. For each Module students are expected to complete an experiment, submit a Post-Lab Workbook, and take a graded quiz. A midterm and a final are given to the students that covers safety information as well as the content presented within each Module. To ensure that students feel confident in their ability to complete the experiment each week, TAs host office hours that are spread throughout the workday every Monday through Friday.

The Post-Lab Workbooks contain pre- and post-lab questions as well as data tables for students to fill out. In addition to analyzing their data, students are also required to document themselves completing the lab by taking pictures of the lab. Students are directed when to take the pictures and TAs will give students feedback on their technique. For example, students are required to assemble a filtering flask to separate a sample during their first experiment of the semester. Students must take a picture of their filtering flask set-up before they filter their sample, and the TAs are directed to check the image to make sure students created a seal by wetting their filter paper. It is understood that this level of feedback is hard to give when low resolution images are submitted. It should be noted that not all experiments required students to take a picture that would require TA feedback. Most of the images that were submitted by the students were used to validate the student's personal completion of the experiment. For example, in a different lab, students are required to take an image of their well plate after mixing the required chemicals. The students are graded based on their observations in this experiment and it would be easy to fudge to observations without completing the experiment. The requirement of this picture serves to confirm the student completed the experiment and would only require a comment if there was a serious safety violation observed.

Oglethorpe University

Traditionally, Oglethorpe University chemistry labs were conducted using 18 person sections with groups of 2 students conducting experiments with guided problem-solving sessions built-into lab time. Students were expected to utilize scientific analysis of data to present their work from the labs. This method was unfeasible after March 2020 due to quarantine restrictions, which led to the filming of experiments by instructors. Lectures were conducted using question-answer problem solving sessions built into class time to facilitate critical thinking skills among students in real-time. Additionally, student instructor (SI) led tutoring sessions assisted in both lecture and lab materials, with approximately 25% of general chemistry students attending extracurricular problem-solving sessions.

Regarding rigor during the in-person laboratory, students focused upon data analysis and presentation regardless of perfection of data acquisition. This rhetoric was reinforced during the online lab sessions, where data was presented to students. Interestingly, the majority of general chemistry students reported that the focus on data analysis during the online labs provided more benefit for developing their statistical analysis skills, as there was less focus on techniques and more focus on theoretical analysis.

During the Fall 2020 semester, research presentations were added as a course requirement. Students presented two different presentations to their peers that would be graded upon content and their presentation skills. For the first presentation, students were allowed to pick an atom and explain it to their classmates. For the second presentation, students allowed to pick between the general reaction types to research and explain to their peers. The grade that each student received came from a mix of instructor, self, and peer evaluations. This allowed for rigor to be maintained by utilizing a rudimentary peer-review process (while exposing students to this concept). Some of these projects lead to amazing results as students really gravitated towards the creative aspects of this project. One specific example includes the merging of creative and scientific talents, in which a student produced an animation showing the discovery of Fluorine, with proper scientific references and moving atomic structures.

York Technical College

York Technical College has historically conducted all chemistry courses in an in-person environment. Course sections were capped at 24 students and students worked in pairs to complete lab experiments. Regardless of the section enrolled, all students in a given chemistry course completed the same laboratory experiments, collected the same data, and completed a standardized post-lab worksheet. However, individual instructors were afforded autonomy as it related to the pre-lab lecture, lab activities, and lab quizzes. In some sections, a pre-lab worksheet was implemented to assist in concept understanding, whereas in other sections, students completed activities in the Knewton Alta online platform.

In the Spring 2020 term, all chemistry labs were moved to the online environment after spring break due to the COVID-19 pandemic and subsequent closure of all public schools and colleges in South Carolina. Labs were conducted through recorded demonstrations and technique videos for the remainder of the Spring 2020 term combined with post-lab questions; sample data was provided as necessary. At the same time, chemistry faculty created lab activities using PhET and ChemCollective simulations to implement in subsequent terms.

During the Summer 2020 and Fall 2020 terms, pre-lab lecture was conducted via either a live or recorded session with the instructor. In these terms, the pre-lab lecture was frequently used to provide students an additional opportunity for mastery of concepts and calculations. The majority of labs were conducted either via demonstration with data provided for analysis or through student exploration of concepts using PhET or ChemCollective simulations, while other labs were conducted.

Evolution of Rigor

UAB Changes Made after Spring 2020

At the end of the spring term, students in both of the author's lectures were asked to help design their ideal remote course. A survey was administered via Canvas that polled students on delivery method, course setup and desired spacing of assessments. These results were used to guide the development of the Summer 2020 offering of general chemistry II. Students indicated that they would prefer directed reading prepared by the instructor directly in Canvas and to be held accountable for preparing for class. They also indicated that online assessments were preferable and that, if they knew in advance, were willing to be live proctored in trade for extra time on exams.

For the Summer 2020 offering of general chemistry II, course specific text was built directly in Canvas. Modules were utilized to present material roughly one chapter at a time. These chapters were subdivided into smaller sections, each with their own page in Canvas. Each page was set up beginning with objectives and a note-taking worksheet, which guided them through taking notes on the content that followed. These worksheets also included note-taking prompts, in the form of questions students should answer or "things to think about" type prompts, pertaining to images and videos. The content that followed the objectives was largely text and included the necessary background and extra practice to succeed on the exam. Where appropriate, examples were worked out first in the text with detailed explanation with additional recorded examples following. Videos for the summer course were also voice over PowerPoint, as that was a quick and easy way to make both extra practice videos while developing a great deal of content in Canvas. At the end of each section page was a reading quiz designed to ensure students had reviewed the material in the content page. These quizzes are mostly definition type questions, but very simple calculations are included where appropriate. Students could earn bonus points at the end of the term if their average on these quizzes was high enough.

Graded assessments during the summer took the form of lower stakes "Module" quizzes, which tested students over material one chapter at a time. These quizzes were not proctored, nor did they require the use of a lockdown browser. The summer course also required students to take a high-stakes midterm and final exam. These two exams were administered via Canvas with use of ProctorU live tutoring and were each open for 3 days. With the exception of students registered with the University's Disability Support Services office, all students were required to use the ProctorU service. This made the high stakes assessment more secure. The course design for the fall 2020 and spring 2021 semesters was very similar, though the number of graded assessments were adjusted.

Due to space constraints and a consensus among department faculty that the organic sequence be given priority for on campus lab offerings, laboratory courses in the General chemistry sequence remained remote for the Fall 2020 and Spring 2021 semesters. Lab sections were each 170 minutes. TAs held office hours for the first hour of that time, giving the students a chance to ask questions individually. Students logged into Zoom 1 hour after the scheduled lab time, heard a brief introductory lecture from TAs, and worked as groups in breakout rooms for the remainder of the time. The focus of the lab content shifted from hands-on training to other skills scientists need to develop, such as robust notebook keeping, developing a command of the scientific literature, and working with collected data. Activities in both of these semesters varied widely depending on the content. Some of the videos prepared by TAs for the Spring 2020 term were still utilized, as were a number of simulations from various sources. Activities were even created in collaboration with other campus offices, such as the assignment on scientific literature developed by a campus librarian. A strong emphasis was placed on developing Excel skills, which has been praised by our Engineering department. The final, high stakes requirement for the lab was to present a poster or brief presentation as a group.

Oglethorpe Evolution of Rigor

For the summer semester of general chemistry 1 and 2 labs, Oglethorpe students were expected to complete guided data analysis of data from previous labs (data was taken during the previous spring semester, 2020). While this method proved adequate, it deducted the need for active learning and participation from students. Regardless, student performance mirrored that of previous labs, with a minimal digression in grade averages.

This method was thoroughly improved upon during the fall 2020 semester, which saw the use of both laboratory demonstrations (via video) along with tailored simulation experiences. Online simulations appeared to be both desirable accessible to students, in comparison to raw data calculations. Intermingled with the simulations were student-led research presentations about molecules, elements, and reactions. These exercises were implemented to promote presentation skills and research ethics to new science majors (these classes were predominately attended by first semester freshmen students). The Instructors also included a few "live" demonstrations via video, with the intention of allowing the students to visualize the laboratory space in which experiments were being conducted. Rigor was maintained during these events by educating students on proper notebook etiquette, along with graded full notebook entries outline the fundamentals of data collection and presentation.

With the spring 2021 semester being predominantly face-to-face for labs, only one section of online general chemistry labs remains at Oglethorpe. This section keeps to the previous semester's rigor of simulations with video demonstrations, but with more emphasis on the video lab experiments and proper lab notebook dictation. This method allows for easy transition of face-to-face students to online modality (and back) without excessive disruption in pedagogy. Indeed, the student accountability for transitioning between face-to-face and remote learning has proven easier utilizing this method.

York Tech Evolution of Rigor

As the pandemic progressed and the need for online asynchronous and synchronous courses remained apparent, chemistry instructors worked to ensure an adequate learning experience for students. They continued to deliver virtual live-lectures and provided recordings to students who were unable to attend. Instructors also incorporated new techniques to engage students during the class meeting times through problem solving activities and the use of breakout rooms and polls during live-streamed lectures. Instructors across the college and especially in the science disciplines are frequently encouraged to add low-stake formative assessments to their courses. Increased consistency and rigor were evident, especially with the laboratory component of general chemistry I and II (CHM 110/111).

Customized laboratory kits from Carolina Biological were implemented in general chemistry I and II courses that were delivered in either an asynchronous or synchronous online modality in the Spring 2021 term. In general chemistry I and II, three hands-on experiments were combined with simulations and demonstrations to ensure development of specific laboratory skills and to ensure appropriate topics were covered.

Hybrid courses were introduced for general chemistry I and II and the GOB chemistry course in the Spring 2021 term, providing an online environment for lecture, and a few in-person lab opportunities complimented by simulations. In order to maintain social distancing in the laboratory environment, students in hybrid courses were restricted to three in-person labs, with the remaining labs being completed using simulations. Students in hybrid chemistry I courses completed measurement, significant figures, mol ratio, formula of a hydrate, and chemical reactions laboratory experiments in person. Students taking either synchronous or asynchronous online chemistry I courses received physical lab activities at no charge for three topics: density, chemical reactions, and freezing point depression. The differences in hands-on labs were necessary due to shipping timeframes for the lab kits. In chemistry II, students in hybrid courses completed Le Chatlier's Principle, concentration of an acid - including buffers, and entropy lab activities in person. Those in an asynchronous or synchronous online chemistry II course received lab experiment materials for reaction rates, buffered solutions, and oxidation - reduction.

Closing Statements

In this chapter we have discussed how rigor can be maintained in an online chemistry course, specifically through course development, delivery, and assessment strategies. A summary comparing the different institutions can be found in Table 1. For content delivery, maintaining rigor was found to be centered around not only the topics covered, but also how the content was delivered to the students. It is important to recognize that students in an online class should be exposed to the same topics that are covered in a traditional in-person classroom. The inclusion of all topics in an online course will prevent students from opting into an online course simply because they view it as easier or "watered-down".

Content covered in online courses should not only match the content covered in a traditional course, but it should also be developed to ensure students can master course learning outcomes. There are several different ways that the material can be presented to the students (pre-recorded videos, voice-over PowerPoints, synchronous lectures, asynchronous independent study) and it is the instructor's responsibility to make sure that they select the appropriate delivery method for the level of their course and the composition of their students. The authors have a particular preference for making short, pre-recorded videos covering specific topics, as this is a known best practice for teaching online, and encourage other instructors to incorporate them into their online course whenever possible (*18*, *19*). Adding these videos to an online course will give students an immediate reference to any topic they would like to review.

A common theme emerged from all institutions involved in this paper when assessment strategies were evaluated. It became apparent that having a proctoring system for students while taking an exam is an important component to maintaining rigor in an online classroom. In this paper, three different proctoring strategies were used: ProctorU, HonorLock, and Zoom. Another common theme was the addition of low-stakes quizzes. To keep students on pace, it is advised that formative assessments be incorporated into the online class. This will provide students with intermittent feedback about their progress on specific topics and will allow them to realize their weakness before taking a major summative assessment.

For courses with a laboratory component, multiple avenues of instruction can be utilized to ensure students receive an acceptable lab experience. The institutions herein employed the use of demonstrations, data sharing, pre- and post- lab assignments, and physical laboratory kits for home use to provide students as much of a laboratory experience as possible. While simulations provide appropriate/ample opportunity for students to visualize chemistry concepts, the development of laboratory skills can only be accomplished through hands-on completion of lab experiments. In an online chemistry course, these skills can be developed through the use of physical lab kits shipped directly to students. The first weeks of subsequent courses may be spent learning techniques that would have been mastered by those students who were fully remote but were not allotted laboratory kits. An alternative to this would be to hold a "bootcamp" style workshop where students can practice missed skills prior to enrolling in more advanced laboratory courses, such as the newly developed "ACS Hands-On Lab Skills Course for Students".

The application of academic rigor in online courses can be molded to fit the needs of any institution, instructor, or course. As described above and outlined in Table 1, there are multiple ways to ensure rigor is maintained in an online environment. All courses and institutions described herein ensured that courses covered all necessary content and that students were assessed appropriately, regardless of course delivery method. Additionally, the instructors and departments included adapted course materials, content delivery methods, and laboratory activities to ensure continued rigor in online chemistry courses. As described in the text above, there are many ways in which to ensure rigor in an online chemistry course, with no absolute correct answer. The authors believe that by sharing their experiences regarding maintaining rigor during a rapid transition to teaching online, and the evolution of their courses in the subsequent semesters, the reader can realize that keeping rigor in an online classroom is both possible and important.

Table 1. Summary of Rigor at Three Different Institutions

Applications of Rigor		UAB	Oglethorpe	York Tech
Content Presentation	Spring 2020	Developed content videos for students to watch before coming to class; Used synchronous lecture time as problem-solving sessions	Synchronous class sessions in a lecture/ example/ small group practice.	Synchronous sessions with recordings for students to review.
	Evolution of Rigor	Added custom course content to replace the traditional textbook	Increased student participation via student explanations during class	Added active learning activities into synchronous sessions
Assessment	Spring 2020	Shifted traditional exams online; used ProctorU for remote proctoring.	Used Zoom to proctor exams; theoretical molecules	Moved traditional exams into the LMS; updated test settings & Honorlock remote proctoring.
	Evolution of Rigor	Added more formative assessments (daily quizzes)	Focused more on "work shown" opposed to correct answer in grading	Added low-stakes formative assessments; use Honorlock remote proctoring.
Lab	Spring 2020	Presented TA-created videos of experiments and had students work in groups to parse out data and analyze it; low-stakes weekly quizzes that built up to the final exam	Guided data analysis using experimental data from a previous term.	Lab demonstrations with data provided; dry labs with sample technique videos and post-lab questions.
	Evolution of Rigor	Added more simulations and hands-on experiments that students were able to complete with common household items	Video demonstrations and simulations. Increased focus on notebook dictation. Introduction of peer-review for presentations.	Interactive simulations, some in person labs, and physical lab kits from Carolina Biological added.

References

1. Holme, T. Introduction to the Journal of Chemical Education Special Issue on Insights Gained While Teaching Chemistry in the Time of COVID-19. *J. Chem. Educ.* **2020**, 97 (9), 2375–2377.
2. *From Disruption to Recovery.* https://en.unesco.org/covid19/educationresponse (accessed March 31, 2021).
3. *The Difference Between Emergency Remote Teaching and Online Learning.* https://er.educause.edu/articles/2020/3/the-difference-between-emergency-remote-teaching-and-online-learning (accessed March 31, 2021).
4. Holme, T. A. Chemistry Education in Times of Disruption and the Times That Lie Beyond. *J. Chem. Educ.* **2020**, 97 (5), 1219–1220.
5. Simon, L. E.; Genova, L. E.; Kloepper, M. L. O.; Kloepper, K. D. Learning Postdisruption: Lessons from Students in a Fully Online Nonmajors Laboratory Course. *J. Chem. Educ.* **2020**, 97 (9), 2430–2438.
6. Koksal, I. *The Rise of Online Learning.* https://www.forbes.com/sites/ilkerkoksal/2020/05/02/the-rise-of-online-learning/?sh=5b12e06172f3 (accessed March 31, 2021).
7. National Center for Education Statistics. *Digest of Education Statistics.* https://nces.ed.gov/programs/digest/d18/tables/dt18_311.22.asp (accessed March 31, 2021)
8. Vesely, P.; Bloom, L.; Sherlock, J. Key elements of building online community: Comparing faculty and student perceptions. *J. Online Learn. Teach.* **2007**, 3 (3), 234–246.
9. Gelles, L. A.; Lord, S. M.; Hoople, G. D.; Chen, D. A.; Mejia, J. A. Compassionate flexibility and self-discipline: Student adaptation to emergency remote teaching in an integrated engineering energy course during COVID-19. *Educ. Sci.* **2020**, 10 (11), 304.
10. Mupinga, D. M.; Nora, R. T.; Yaw, D. C. The learning styles, expectations, and needs of online students. *College Teach.* **2006**, 54 (1), 185–189.
11. O'Carroll, I. P.; Buck, M. R.; Durkin, D. P; Farrell, W. S. With Anchors Aweigh, Synchronous Instruction Preferred by Naval Academy Instructors in Small Undergraduate Chemistry Classes. *J. Chem. Educ.* **2020**, 97 (9), 2383–2388.
12. Stowe, R. L.; Esselman, B. J.; Ralph, V. R.; Ellison, A. J.; Martell, J. D.; DeGlopper, K. S.; Schwarz, C. E. Impact of Maintaining Assessment Emphasis on Three-Dimensional Learning as Organic Chemistry Moved Online. *J. Chem. Educ.* **2020**, 97 (9), 2408–2420.
13. Pellegrino, J. W., Chudowsky, N., Glaser, R., Eds. *Knowing What Students Know: The Science and Design of Educational Assessment*; National Academy Press: 2001. http://www.nap.edu.
14. Cooper, M. M.; Kouyoumdjian, H.; Underwood, S. M. Investigating Students' Reasoning about Acid-Base Reactions. *J. Chem. Educ.* **2016**, 93 (10), 1703–1712.
15. Perets, E. A.; Chabeda, D.; Gong, A. Z.; Huang, X.; Fung, T. S.; Ng, K. Y.; Bathgate, M.; Yan, E. C. Y. Impact of the Emergency Transition to Remote Teaching on Student Engagement in a Non-STEM Undergraduate Chemistry Course in the Time of COVID-19. *J. Chem. Educ.* **2020**, 97 (9), 2439–2447.
16. Newton, D. *Another Problem with Shifting Education Online: Cheating.* https://hechingerreport.org/another-problem-with-shifting-education-online-cheating/ (accessed March 30, 2021).

17. Perlman, L. *With Classes Online, a Wave of Cheating is Ravaging Penn's Academics*. https://www.thedp.com/article/2020/10/cheating-online-semester-penn-covid (accessed March 30, 2021).
18. Hsin, W.-J.; Cigas, J. Short videos improve student learning in online education. *J. Computing Sciences in Colleges* **2013**, *28* (5), 253–259.
19. Guo, P.; Kim, J.; Rubin, R. How Video Production Affects Student Engagement: An Empirical Study of MOOC Videos. *Proceedings of the first ACM conference on Learning @ scale conference (L@S 2014)*. 2014. http://up.csail.mit.edu/other-pubs/las2014-pguo-engagement.pdf (accessed March 31, 2021).

Chapter 6

A How-To Guide for Making Online Pre-laboratory Lightboard Videos

Timothy R. Corkish,* Max L. Davidson, Christian T. Haakansson, Ryan E. Lopez, Peter D. Watson, and Dino Spagnoli*

School of Molecular Sciences, The University of Western Australia, Perth, Western Australia 6009, Australia

*Email: timothy.corkish@research.uwa.edu.au

*Email: dino.spagnoli@uwa.edu.au

This chapter details the creation and development of lightboard videos as an online pre-laboratory resource for first-year undergraduate chemistry students. The lightboard is a transparent whiteboard that allows students to always see the instructor's face while content is delivered, removing the need to turn away from the audience as per a traditional whiteboard video. A series of videos have been created to move content originally used in the pre-laboratory discussion online. The videos focus on typical first-year stumbling blocks; in particular, those that deal with mathematical concepts essential for chemical knowledge. This chapter describes the process of making the online pre-laboratory videos from conception to completion: from multimedia design considerations to equipment specifications and setup, as well as some useful tips for filming. Student responses to the videos are also covered in brief, including some suggestions for future topics. Ultimately, this chapter aims to serve as a resource for those looking to incorporate lightboard technology as an online pre-laboratory video within an undergraduate chemistry course.

Introduction

What Is a Lightboard?

Universities worldwide are increasing their efforts to provide high-quality online courses. The main driver for online courses is to increase the number of students, and therefore revenue, without incurring the cost of new building infrastructure (1, 2). The need for high-quality online courses has been exacerbated and accelerated due to the COVID-19 pandemic that affected the globe in 2020. Lightboard technology is a growing area of innovation in education which can provide high-quality engaging online videos (3). The first lightboard used in a higher education setting was developed in

2013 by Michael A. Peshkin, Professor of Mechanical Engineering at Northwestern University, who sought to present lecture material in a flipped or blended approach (4). A lightboard is a piece of glass that has a lower iron content than typical glass, allowing more light to pass through the glass and removing the green hue that is present in glass with higher iron content. Bright white LEDs light up the internal cross section of the glass and the light travels through the glass by total internal reflection (TIR). Fluorescent, non-permanent markers are used on the glass and the marks glow brightly due to their illumination from frustrated total internal reflection (FTIR). The contrast of the marks is even more pronounced in a low light environment and if a black backdrop is used as the background to the video recording, as shown in Figure 1. The presenter stands on one side of the glass and the video camera is on the opposite side (see the Equipment and Setup section for more information). Thus, during the editing of the video, the recording needs to be reflected about the y-axis so that the text and diagrams are correct for the viewer (i.e., a right-handed presenter would appear left-handed).

Figure 1. A screenshot of a lightboard video detailing a series of titration calculations. Note that the instructor appears to be drawing left-handed when they are actually right-handed.

In chemistry tertiary education, there have been some reports of the use of lightboards (5–7). Skibinski and co-workers used a lightboard projection system in the lecture hall, concluding that the advantages of this particular set up were that the lecturer faced the students and that it was inexpensive compared to large-screen tablets with a stylus (5). They also commented that a lecturer's handwriting can degrade with a stylus and tablet (5). Lightboard videos also lend themselves to a flipped or blended classroom approach as pre-lecture material. Fung concluded that there are advantages for the academics due to ease of creation compared to a PowerPoint voice-over video, and that video production utilizes very user friendly, accessible editing tools (6). Lightboard videos showing reaction mechanisms that supplement the traditional mode of delivery (face-to-face lectures) have been shown to increase student engagement with the subject material (7). The majority of previous studies, highlighted above, have focused on using the lightboard technology as an online video to supplement face-to-face classes, such as lectures. Lightboard technology can also be used to 'flip' the pre-laboratory discussion and therefore act as a means of preparing students for the laboratory. This chapter will discuss how lightboard technology can be used to make effective pre-laboratory videos to enhance student learning.

Online Pre-laboratory Activities

The purpose of a pre-laboratory activity (online or in-class) is to prepare students to learn and understand a new topic. In a review of 45 years of literature on pre-laboratory activities (8), Agustian and Seery framed the laboratory as a complex learning environment (9). Based on this framework, Agustian and Seery proposed that the information in a pre-laboratory can be supportive (core material needed to understand why approaches are being taken) or procedural (material needed to carry out the experiment). Supportive information tends to be of higher value to students if presented before the laboratory. Procedural information is better suited if presented as the student requires (8). However, a study by Teo et al. found that students were encouraged to learn independently in the laboratory and maintained higher levels of cognitive engagement after watching online instructor-narrated demonstration videos of the practical (procedural information) accompanied by pre-laboratory questions for discussion (supportive information) (10).

A supportive or procedural online pre-laboratory activity can be very effective for student laboratory learning, especially in the affective domain of learning. Students who have engaged or watched online pre-laboratory activities feel more prepared for the laboratory (11–14), are more motivated to learn (15, 16), and demonstrate an enhanced laboratory learning experience (17, 18). Moreover, there are recent studies to suggest that an online pre-laboratory activity can improve student learning outcomes, with the pre-laboratory itself allowing for a greater flexibility in students learning at a self-directed pace, thereby managing cognitive load and encouraging students to take responsibility for their own learning (13, 19, 20). Additionally, the online pre-laboratory can lead to an improved experimental practice, resulting in more time in the laboratory for conceptual learning (13, 19, 20).

As described in a study by Loveys and Riggs, there are two 'rules of thumb' to consider when using a video as part of a pre-laboratory activity: keep it short, and humanize it (19). These two considerations can be easily integrated into a lightboard video. Videos can be kept short and concise through the preparation phase and editing process. The activities can be humanized by the instructor always being visible to the student.

Not all pre-laboratory content is adaptable to the style of a lightboard video. For example, a video which needs to include the procedural information of an experimental technique would not be suited to a lightboard video. The types of pre-laboratory videos detailed in this chapter have focused on the mathematics needed for students to complete the analysis of data. Student mathematics proficiency has been known for some time as an important contributor, or even an indicator, of their effectiveness as a chemist (21–23).

Video Design

As an introductory note, links to the produced lightboard videos have been provided in references (24–27) (24–27). The standard format adopted for lightboard videos was as follows. The aim was for videos to be 8-10 minutes long, although some concepts were more complex and therefore required more time for explanation. The first 2-3 minutes of the video are used to present learning outcomes and key concepts prior to the worked lightboard example. This introduction is referred to as a 'talking head' section, with PowerPoint slides presented along with a green screen cut-out of the instructor as shown in Figure 2. The remainder of the video features the worked lightboard example, broken up across multiple lightboard scenes in order to appropriately segment the example.

Figure 2. A screenshot of the 'talking head' introductory portion of the video.

Mayer explains in his Cognitive Theory of Multimedia Learning that humans have two competing, and limited, processing channels within the context of a multimedia lesson: auditory and visual (28). Multimedia learning then has unique challenges with respect to cognitive processing. For example, the selection and integration of relevant words or graphics within that multimedia lesson. Mayer argues that multimedia lessons should be designed with this in mind, with a careful onus on not overloading the limited processing channels, thereby leading to a greater likelihood of effective learning occurring. The remainder of this section will introduce several principles of multimedia design put forward by Mayer (29), and discuss how lightboard videos should be designed to adhere to them. In particular, lightboard videos adhere most effectively to the dynamic drawing and gaze guidance principles.

Dynamic Drawing Principle

This principle states that in an instructional video, people tend to learn more effectively when the instructor draws while they speak, as opposed to addressing static images (3, 29). Mayer even goes on to state that videos should include images of the presenter's hand drawing, in contrast to a Khan Academy style video, for example, where although nothing is pre-drawn, no hand or person is shown writing or drawing (3, 30).

Lightboard videos, by their very nature, therefore do adhere to the dynamic drawing principle. The instructor's hand is always shown onscreen during the drawing of text or figures. Additionally, due to the transparency of the glass, the instructor does not obstruct the drawn content as it is being presented. It has been found to be useful from the perspective of the presenter to pre-draw some content in order to scaffold the shot (see Figure 3), although it remains to be seen if this detracts from the learning experience as per the dynamic drawing principle. The author's advice here would be to use the signalling principle of multimedia design, discussed in further detail below, to draw attention to any essential pre-drawn or pre-written content (29).

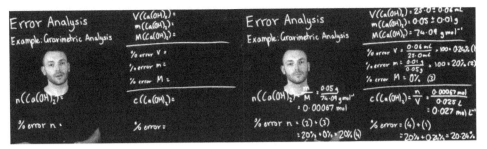

Figure 3. Image of the lightboard before (left) and after (right) working through an error analysis example. Pre-drawn content shown in the left image is useful in scaffolding the worked example. Pre-drawn content was found to be helpful for both student and presenter.

Gaze Guidance Principle

The gaze of the instructor is an important component of any video as it can direct the viewers' attention and influence their engagement with the content. Mayer states through the gaze guidance principle of multimedia design that learning is improved when the instructor can switch their gaze between the camera and the written content (30). Lightboards in particular are an effective way to manage the gaze guidance of the presenter. Firstly, students watching the video can have direct eye contact with the presenter, which is believed to build a social relationship and encourages the student to learn (31). Secondly, the presenter can shift their gaze over to important content, which helps to direct the student's own gaze. An effective lightboard video should be recorded with this design principle in mind.

Segmenting Principle

The segmenting principle states that learning is more effective when the presented content is split into several smaller components, whereby the pacing can be controlled by the user (32). Both the segmenting principle as well as the pre-training principle, to be outlined in detail later, are discussed in reference to minimizing the essential processing aspect of cognitive load; that is, processing due to the presented material (32). This is separate to extraneous processing, which is the processing of material not directly tied to the overall learning outcomes.

There are several different approaches that can be used to segment lightboard videos. The first method involves splitting content across different lightboard takes or scenes. Figure 4 illustrates this by comparing the same drawn example in two scenarios: the first in only one lightboard scene and the second spread across two scenes. In addition to segmenting the content, the examples themselves become far clearer when spread over two different scenes, provided the segmenting is done at appropriate points. As a general piece of advice, distilling each lightboard scene into one key idea or calculation would aid in enacting the segmenting principle within the video.

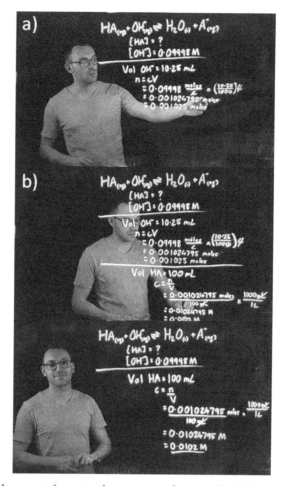

Figure 4. Contrast between information being presented on a single lightboard (a) versus the same information spread across multiple lightboards (b).

Another way to segment the content of lightboard videos is to spread the separate takes over different videos altogether, creating a user-paced element. As an extension of this, if an entire concept becomes too large to tackle in just one video, consider segmenting it into further videos.

Pre-training Principle

This principle dictates that learning is more efficient when a student is exposed to key terms and concepts prior to the main video (*32*). According to the pre-training principle, key terms should be introduced prior to the main content rather than presented concurrently with the worked examples. Introducing key concepts at the same time as worked examples can lead to some confusion or cognitive overload (*32*).

As per the constructive alignment approach (*33*), in which learning outcomes, activities, and assessments all intertwine within a taught course, the learning outcomes should be a feature of the beginning of every video. It has been noted in the rationale for constructively aligning a chemistry course that students can see the value in an activity if they know the learning outcomes of the lesson will be matched with what they are to be assessed on in class (*34*). The authors utilized the 'talking head' approach (Figure 2) to introduce key terms and learning outcomes as per the pre-training principle. For the introduction to error analysis lightboard video, for example, this included a small discussion on why error analysis is important in addition to introducing the basic error analysis rules for examples involving addition, subtraction, multiplication and division.

Signalling Principle

The signalling principle (29), as alluded to earlier, states that students learn more effectively when essential information is highlighted or has attention drawn to it. This could be in the form of arrows, titles, or even a vocal cue and hand gesture, and is important to consider during a worked lightboard example.

Subtitling Principle

The subtitling principle states that for students watching a video in their second language, learning is more effective when words are printed in addition to, or even in place of, being spoken (30). If multimedia lessons are uploaded to YouTube, for example, closed captions can be easily embedded into the video. The authors advise not to rely on auto generated captions due to errors in captioning. Always edit the captions that accompany instructional videos to ensure accurate information is being presented. On average, it takes approximately twice the length of the video to correct the automatically generated subtitles e. g. for a ten minute video, the subtitles would require twenty minutes of editing.

Generative Activity Principle

Lastly, the generative activity principle is one to be taken into consideration. This principle describes that students learn more effectively from a video when they are tasked with a learning activity during a lesson, such as taking notes or pausing to attempt an example (30). This principle will be incorporated into future lightboard videos and works seamlessly well with the segmenting principle to include a user-paced element into the learning experience. Moreover, as technology advances there will be scope to integrate online polling questions into the lightboard video to make them interactive. Such interpolated testing has been shown to enhance student learning as well as their integration of information, both from within the same lecture segment as well as from different lecture segments (35).

Equipment and Setup

A summary of the equipment and setup used to record videos is described below, with further specifications and costs discussed in full, later in this section. The presenter stands behind the piece of lightboard glass with the camera positioned to record from the opposite side (see Figure 5). A black canvas is placed behind the presenter to be used as a back drop, with a set of two extra lights either side of the lightboard that are directed towards the presenter. A lapel microphone is worn by the presenter, who also makes use of a set of fluorescent dye markers to write on the lightboard. It is essential to wear clothing that has no writing or images, otherwise these will be mirrored in post-production, which will detract from the content being presented.

Figure 5. Image of equipment and set up for filming lightboard videos. Image (a) shows the presenter behind the lightboard and the camera operator on the opposite side. Image (b) shows the point of view of the presenter behind the lightboard.

The Lightboard

The dimensions of the lightboard detailed in the current videos are approximately 172 cm by 83 cm. A strip of LED lights surrounds the lightboard with an attached rheostat to allow for an adjustable light intensity. An electric height adjustable table is used to support the lightboard and helps to frame the presenter of the video depending on their height. While lightboards are commercially available, retailing for approximately $6000 USD depending on the design and dimensions, the model used for our chemistry videos was built in-house by Professor Paul Bergey at the University of Western Australia (paul.bergey@uwa.edu.au). It is important to note here that smaller versions of the lightboard exist, approximately 115 cm by 65 cm. These lightboards can fit on a tabletop and are ideal for small office spaces; placed in between an office chair and a webcam, for example. The smaller lightboards retail for approximately $2000 USD.

Lights and Camera

The following advice on how to set the scene is provided by our helpful Educational Technology team from the UWA Educational Enhancement Unit. A camera with manual settings was used to record the videos, although if need be, a modern smartphone camera could also be used. The guiding principle behind the manual camera settings used is to brighten up the presenter and writing on the board whilst maintaining as dark a background as possible. As such, the camera settings used are as follows:

- Focus is set to Auto. Most digital cameras will have an autofocus setting to ensure the subject is in focus at all times throughout filming. Manual focus can also be used but is not recommended.
- Shutter speed is set to 1/50. A lower value is better for this setting as it dictates how much light is received by the sensor.

- Exposure Value (EV) is set to be between -1.0 and -1.5. This setting combines both the aperture (or f-number) and shutter speed and helps with the brightness of the image. A negative value avoids the potential of over-exposure in a dark environment.
- Gain is set to Auto. This setting should be adjusted as need be to allow for more light on the presenter.

In addition to the camera settings, it is important that the presenter knows what space is available to write on. To this end, it is recommended to draw a few points of reference that are just out of the camera's field of view of the lightboard, so that the presenter is aware of how much of the board is visible to the audience. These points of reference are visible on the lightboard in Figure 5.

As mentioned in brief, two sources of light have been used for the lightboard videos. The first is the strip of LED lights surrounding the edge of the lightboard, and the second is a pair of standing LED lights either side of the presenter (see Figure 5). The second pair of lights are used to help brighten the presenter's face. A set of clear filters are used over the standing lights to disperse the light source in order to provide an appropriate level of brightness on the presenter.

Software and Editing

The Camtasia software package was used to edit raw footage, although Adobe Premiere Pro and Adobe Premiere Rush have been suggested as alternative software. Final Cut Pro and iMovie are also potential options that can be utilized by Mac users. The lightboard videos are hosted on YouTube as this was found to be helpful in embedding the videos on the university's online learning platform. The YouTube platform also includes a range of analytics features. These features, among other things, allow the video uploader to see when the video was accessed and the average duration the video had been watched. As a side note, the videos can remain unlisted on YouTube. An unlisted video is preferred over a private video setting on YouTube. All students email addresses in a course must be listed on YouTube to view a private video. This is not practical in most large first year chemistry classes. Students only require the YouTube URL to be able to access an unlisted video. However, unlisted videos will not appear in any YouTube or Google search. This is useful if it is desirable to know if students are viewing the video, to aid in gathering usage analytics for specific research studies.

Budget

Table 1 documents the approximate costs of setting up a lightboard recording studio based on the equipment used by the authors, as well as some suggestions. The two subtotals in Table 1 represent the costs associated with two different lightboard approaches: the first making use of a larger lightboard, digital camera with manual settings, and extra lighting, with the second approach being more conservative with a tabletop lightboard and a webcam.

The price of $100 USD for a lapel microphone is an estimate based on some available wireless models. This was found to be the most effective way of recording audio, especially as a second person can monitor the audio with a set of headphones plugged into the camera, with the wireless audio receiver also plugged in. Other lapel microphones that plug in directly to a smartphone are a handy alternative option and can be purchased for less than $40 USD.

Table 1. Approximate costs for the setup of a lightboard recording studio. A more detailed approach is included, as well as that requiring only a smaller lightboard and a webcam as essential items.

	Item	*Approximate Cost ($USD)*
Approach 1	Lightboard (173 cm by 82 cm)	6000
	Camera	750
	LED Lights	100
	Markers	10
	Lapel Microphone	100
	Black Canvas	40
	TOTAL	**7000**
Approach 2	Smaller Lightboard (115 cm by 65 cm)	2000
	Webcam	100
	Markers	10
	Black Canvas	40
	Lapel Microphone	40
	TOTAL	**2190**

Tips for Filming

The following constitutes some useful tips to consider during the filming and editing processes:

Preparation

- While it is possible to record the lightboard videos detailed in this chapter with just one person, a team of two or more works best. This second person can act as the director: counting the presenter in, checking the clarity of the audio recording, and generally advising on the recorded take. It is important to note that the team that developed and created the lightboard videos in the provided references (24–27) have no extensive video creating or editing experience. Therefore, this approach is suitable for novices.
- A sheet of notes just offscreen can help the presenter deliver information. Worked mathematical examples that feature multiple lines can be difficult to remember for the presenter. Ironically, it must be remembered that the presenter's own cognitive load should be managed during filming. The challenge of speaking, writing, and gesturing on the fly can be eased with a sheet of notes detailing the content to be written on the lightboard. This has proven most useful for quick answers to the mathematical examples, especially those with a considerable number of significant figures.
- Certain topics may need the consideration of careful planning. Chirality as a topic in chemistry, for example, could be affected by the inversion of the filmed lightboard during editing (6). Another example may be the right-hand rule of physics. These subjects may

lend themselves to an explanation through a different format, or else require careful planning to make them work for presentation through a lightboard video.

Pre-drawn Items

- Both lightboard and whiteboard videos allow students to follow at the same pace as the presenter's writing. Despite this, it can be useful to have some items pre-drawn before recording. As mentioned previously, this can seem counterintuitive as per the dynamic drawing principle of multimedia design. However, there are certain elements that can be helpful to draw before recording in order to create a scaffold. Tables, information from an earlier lightboard scene, such as a value from a previous worked example, and certain placeholders all fall into this category. For example, if recording a lightboard video regarding titrations, it could be a good idea to have the overall reaction as a figurehead during every take (see Figure 1). Not only does this allow for attention to be drawn to the item as per the signalling principle, it also helps to frame the space available during the entire take.
- As a continuation from the last point: make use of the reverse side of the board. That is to say, the side not being written on during filming. It can be strenuous to constantly erase drawn information in between takes. Pre-drawing a table, for example, on the reverse side of the board means that the same table will not have to be erased and re-drawn each new take. Only the table outline in this case would have to be pre-drawn.

The Presenter

- It has been noted that lightboard videos face their own challenge with respect to directing the student's attention through the presenter (36). In other words, the presenter has the potential to act as a distraction to the content being drawn at certain times. Being aware of the presenter's eye contact and visual cues, such as looking to a particular section of information on the lightboard, can be an important tool in making an appropriately engaging video, as per the aforementioned gaze guidance principle. A similar philosophy should extend to hand gestures and body language, in that the presenter should consider these cues deliberately throughout filming.
- It is important to consider the choice of clothes worn during filming. Darker clothing works best because the presenter may be standing behind words and diagrams. Moreover, there should be no logos, text, or pictures on the clothes as they will be reversed along with the rest of the footage in post-production.

Student Responses

This section constitutes general trends resulting from open-ended survey questions regarding the lightboard videos. The Human Ethics Office at UWA reviewed surveys, consent form and participation information forms and this study was authorized to proceed under project number RA/4/1/8993 for a five-year period. When students were asked if they enjoyed the pre-laboratory videos and asked to comment why or why not, the vast majority of students responded that they did enjoy the videos, specifically mentioning them being clear and informative:

"Yes, the lightboard was an interesting approach compared to standard whiteboard videos."

"I did enjoy it. It was very clear and easy to understand."

"Enjoyed it as the working out was done in a systematic way where items were grouped in colours. I am a visual person so this helped."

"Yes, made me feel confident going into my chemistry lab."

Students were then asked what aspects of the videos need improvement and what changes they would suggest. In a general sense, students commented that they would like to see more examples, including those that are more complex. Some examples of these responses are shown below:

"A few example questions at the end of the video."

"More examples possibly could be implemented to show how to conduct error analysis."

"Possibly including some harder examples."

Lastly, students were asked to suggest which topics in chemistry they felt would benefit from using a lightboard as an online resource to assist in their studies. Many students wrote that any area involving calculations or mathematical equations would benefit from the lightboard approach, similar to the videos they had already experienced. Students also noted area such as kinetics, chemical equilibrium, and intermolecular forces, among others. Interestingly, students also suggested organic chemistry and reaction mechanisms as possible topics for future lightboard videos. Overall, the lightboard videos of this chapter were met with positive reception by the students, and the mathematical concepts outlined within them have been justified by the student responses.

Conclusion

This chapter has described a rationale and procedure for creating pre-laboratory lightboard videos as part of an undergraduate chemistry course. Multimedia design principles have been discussed in relation to their implementation in recording a lightboard video, including the dynamic drawing and gaze guidance principles, to which the lightboard format adheres particularly effectively. In addition, equipment, setup, and costs have been addressed. In setting up a lightboard studio, a smaller, tabletop lightboard can be used on a smaller budget, along with a basic webcam or even a smartphone camera. Advice for filming has also been included as items the authors have found particularly useful in easing the recording process. Students were surveyed with open-ended questions and provided some suggestions for future topics, as well as commenting positively on their overall experience with the videos. Overall, lightboard videos are an emerging alternative in the execution of pre-laboratory content, and this chapter presents itself as a starting resource for those looking to implement lightboard technology within an undergraduate chemistry context.

Acknowledgments

The following are acknowledged for financial support: UWA Faculty of Science, School of Molecular Sciences, Australian Government Research Training Program Scholarship, Aarhus University Centre for Materials and Crystallography and the UWA Dean's Excellence in Science PhD Scholarship. The Educational Enhancement Unit at UWA, in particular, Ms Miela Kolomaznik and Ms Irene Lee, are thanked for their help. Professor Paul Bergey is also thanked for his helpful discussions regarding lightboards.

References

1. Choe, R. C.; Scuric, Z.; Eshkol, E.; Cruser, S.; Arndt, A.; Cox, R.; Toma, S. P.; Shapiro, C.; Levis-Fitzgerald, M.; Barnes, G.; Crosbie, R. H. Student Satisfaction and Learning Outcomes in Asynchronous Online Lecture Videos. *CBE—Life Sciences Education* **2019**, *18* (4), ar55.
2. Seaman, J. E.; Allen, E. I.; Seaman, J. *Grade Increase: Tracking Distance Education in the United States*; Babson Survey Research Group: 2018.
3. Fiorella, L.; Stull, A. T.; Kuhlmann, S.; Mayer, R. E. Instructor Presence in Video Lectures: The Role of Dynamic Drawings, Eye Contact, and Instructor Visibility. *J. Educ. Psychol.* **2018**, *111* (7), 1162–1171.
4. Peshkin, M. A. *Lightboard home*. https://lightboard.info/ (accessed 25th March 2020).
5. Skibinski, E. S.; DeBenedetti, W. J. I.; Ortoll-Bloch, A. G.; Hines, M. A. A Blackboard for the 21st Century: An Inexpensive Light Board Projection System for Classroom Use. *J. Chem. Educ.* **2015**, *92* (10), 1754–1756.
6. Fung, F. M. Adopting Lightboard for a Chemistry Flipped Classroom To Improve Technology-Enhanced Videos for Better Learner Engagement. *J. Chem. Educ.* **2017**, *94*, 956–959.
7. Schweiker, S. S.; Griggs, B. K.; Levonis, S. M. Engaging Health Student in Learning Organic Chemistry Reaction Mechanisms Using Short and Snappy Lightboard Videos. *J. Chem. Educ.* **2020**, *97* (10), 3867–3871.
8. Agustian, H. Y.; Seery, M. K. Reasserting the role of pre-laboratory activities in chemistry education: a proposed framework for their design. *Chem. Educ. Res. Pract.* **2017**, *18* (4), 518–532.
9. van Merrienboer, J. J. G.; Kirschner, P. A.; Kester, L. Taking the Load Off a Learner's Mind: Instructional Design for Complex Learning. *Educ. Psychol.* **2003**, *38* (1), 5–13.
10. Teo, T. W.; Tan, K. C. D.; Yan, Y. K.; Teo, Y. C.; Yeo, L. W. How flip teaching supports undergraduate chemistry laboratory learning. *Chem. Educ. Res. Pract.* **2014**, *15* (4), 550–567.
11. Jolley, D. F.; Wilson, S. R.; Kelso, C.; O'Brien, G.; Mason, C. E. Analytical Thinking, Analytical Action: Using Prelab Video Demonstrations and e-Quizzes To Improve Undergraduate Preparedness for Analytical Chemistry Practical Classes. *J. Chem. Educ.* **2016**, *93* (11), 1855–1862.
12. Spagnoli, D.; Rummey, C.; Man, N. Y. T.; Wills, S. S.; Clemons, T. D. Designing online pre-laboratory activities for chemistry undergraduate laboratories. In *Teaching Chemistry in Higher Education-A Festschrift in Honour of Professor Tina Overton*; Seery, M. K., McDonnell, C., Ed.; Creathach Press: 2019; pp 315–332.
13. Sarmouk, C.; Ingram, M. J.; Read, C.; Curdy, M. E.; Spall, E.; Farlow, A.; Kristova, P.; Quadir, A.; Maatta, S.; Stephens, J.; Smith, C.; Baker, C.; Patel, B. A. Pre-laboratory online learning resource improves preparedness and performance in pharmaceutical sciences practical classes. *Innov. Educ. Teach. Int.* **2019**, *57* (4), 1–12.
14. Rodgers, T. L.; Cheema, N.; Vasanth, S.; Jamshed, A.; Alfutimie, A.; Scully, P. J. Developing pre-laboratory videos for enhancing student preparedness. *Euro. J. Eng. Educ.* **2020**, *45* (2), 292–304.

15. Moozeh, K.; Farmer, J.; Tihanyi, D.; Nadar, T.; Evans, G. J. A Prelaboratory Framework Toward Integrating Theory and Utility Value with Laboratories: Student Perceptions on Learning and Motivation. *J. Chem. Educ.* **2019**, *96* (8), 1548–1557.
16. Pogacnik, L.; Cigic, B. How To Motivate Students To Study before They Enter the Lab. *J. Chem. Educ.* **2006**, *83* (7), 1094–1098.
17. Spagnoli, D.; Wong, L.; Maisey, S.; Clemons, T. D. Prepare, Do, Review: a model used to reduce the negative feelings towards laboratory classes in an introductory chemistry undergraduate unit. *Chem. Educ. Res. Pract.* **2016**, *18* (1), 26–44.
18. Chaytor, J. L.; Mughalaq, M. A.; Butler, H. Development and Use of Online Prelaboratory Activities in Organic Chemistry To Improve Students' Laboratory Experience. *J. Chem. Educ.* **2017**, *94* (7), 859–866.
19. Loveys, B. R.; Riggs, K. M. Flipping the laboratory: improving student engagement and learning outcomes in second year science courses. *Int. J. Sci. Educ.* **2018**, *41* (1), 64–79.
20. Fleagle, T. R.; Borcherding, N. C.; Harris, J.; Hoffmann, D. S. Application of flipped classroom pedagogy to the human gross anatomy laboratory: Student preferences and learning outcomes. *Anat. Sci. Educ.* **2018**, *11* (4), 385–396.
21. Scott, F. J. Is mathematics to blame? An investigation into high school students' difficulty in performing calculations in chemistry. *Chem. Educ. Res. Pract.* **2012**, *13* (3), 330–336.
22. Weisman, R. L. A mathematics readiness test for prospective chemistry students. *J. Chem. Educ.* **1981**, *58* (7), 564.
23. Shallcross, D. E. A pre-arrival summer school to solve the maths problem in chemistry. In *Teaching Chemistry in Higher Education-A Festschrift in Honour of Professor Tina Overton*; Seery, M. K., McDonnell, C., Ed.; Creathach Press: 2019; pp 77–88.
24. *Introduction to Error Analysis.* https://youtu.be/dLnsSfGDyHA (accessed Feb. 24, 2021).
25. *Calorimetry and Error Propagation.* https://youtu.be/xCdjd7EYejM (accessed Feb. 24, 2021).
26. *Introduction to Dimensional Analysis.* https://youtu.be/h3sLrameYwA (accessed Feb. 24, 2021).
27. *Acid Base Calc.* https://youtu.be/j7WduLI5DDk (accessed Feb. 24, 2021).
28. Mayer, R. E. Cognitive Theory of Multimedia Learning. In *The Cambridge Handbook of Multimedia Learning*, 2nd ed.; Mayer, R. E., Ed.; Cambridge University Press: Cambridge, 2014; pp 43–71.
29. Mayer, R. E. Using multimedia in e-learning. *J. Comput. Assist. Learn.* **2017**, *33* (5), 403–203.
30. Mayer, R. E.; Fiorella, L.; Stull, A. Five ways to increase the effectiveness of instructional video. *Educ. Technol. Res. Dev.* **2020**, *68*, 837–852.
31. Mayer, R. E. Principles based on social cues in multimedia learning: personalization, voice, image, and embodiment principles. In *The Cambridge Handbook of Multimedia Learning*, 2nd ed.; Mayer, R. E., Ed.; Cambridge University Press: Cambridge, 2014; pp 345–368.
32. Mayer, R. E.; Moreno, R. Nine ways to reduce cognitive load in multimedia learning. *Educ. Psychol.* **2003**, *38* (1), 43–52.
33. Biggs, J. Enhancing teaching through constructive alignment. *High Educ.* **1996**, *32*, 347–364.
34. Adams, C. J. A constructively aligned first-year laboratory course. *J. Chem. Educ.* **2020**, *97* (7), 1863–1873.

35. Jing, H. G.; Szpunar, K. K.; Schacter, D. L. Interpolated testing influences focused attention and improves integration of information during a video-recorded lecture. *J. Exp. Psychol. Appl.* **2016**, *22*, 305–318.
36. Stull, A. T.; Fiorella, L.; Mayer, R. E. An eye-tracking analysis of instructor presence in video lectures. *Comput. Human Behav.* **2018**, *88*, 263–272.

Chapter 7

Working It Out: Adapting Group-Based Problem Solving to the Online Environment

J. L. Kiappes[1,2,*] and Sarah F. Jenkinson[3]

[1]Department of Chemistry, University College London, London WC1H 0AJ, United Kingdom
[2]Corpus Christi College, University of Oxford, Oxford OX1 4JF, United Kingdom
[3]Department of Chemistry, University of Oxford, Oxford OX1 3TA, United Kingdom
*Email: j.l.kiappes@ucl.ac.uk

During the pivot to online learning due to the COVID-19 pandemic, much effort focused on the delivery of teaching from instructor to students. However, peer-to-peer learning is key to both intellectual and social engagement in chemistry courses. Both the student teaching and student learning can develop deeper understanding through the process, so maintaining this aspect of our first-year organic course was deemed critical. Here, we describe a series of biological science-oriented organic chemistry workshops that complement the lecture series of the module. After presenting the traditional on-site formulation, we discuss how we have adapted these sessions into both online and hybrid formats. The use of online whiteboards allows for more diverse engagement by students, while the breakout room format presented challenges for instructors seeking to interact in a way that facilitated, rather than disrupted, student dialogue. The method of implementation continues to evolve after each workshop, with multiple possibilities for engaging students in ways that are as inclusive as possible.

Introduction

Organic chemistry is often perceived to be a difficult course (*1*). The subject demands that students become familiar with new vocabulary and concepts. However, beyond simply learning or memorising these individual pieces, they must develop the ability to apply them in a considered and balanced way to understand both how and why chemical reactions of carbon occur. This often places a large cognitive load on the students. Although traditionally a lecture course, strides in engaging students with organic chemistry have been made for some time by recognising the benefits of collaborative learning (*2*): in flipped classrooms (*3*) and in frameworks such as POGIL (process-oriented, guided-inquiry learning) (*4*).

© 2021 American Chemical Society

In our own organic chemistry course, the Oxford tutorial system provides features similar to that of a flipped classroom. While departmental lectures distribute the content, smaller groups within each College solve problems together to apply these concepts with guidance from a tutor. In redesigning the curriculum and format of the course to be tailored to Biochemistry students, we decided to introduce workshops as a third learning environment to provide centralized problem-oriented sessions. Initially designed to be carried out in person, the workshops have subsequently incorporated online tools and applications to reformat them into both online and hybrid formats.

Context

These workshops were originally designed as part of a first-year, two-term module in Organic Chemistry as part of an undergraduate Masters degree in Molecular and Cellular Biochemistry (MBiochem). At the University of Oxford, as at most universities in the United Kingdom, students apply for a place to study a particular subject. All students on the MBiochem degree take the exact same modules at the same time, with a cohort size of approximately 100 students. The Organic Chemistry module (Mechanistic Biochemistry as of the 2020–21 academic year) is traditionally composed of a series of lectures and tutorials. The lectures are provided to the entire cohort as a group, whereas tutorials take place in small groups (typically 2–4 students in a group). For tutorials, students will generally be given a problem to solve and have annotated with feedback in advance of the tutorial. In the tutorial session itself, the students and tutor will discuss the topic generally, often centered around the problem set as well as new "unseen" questions. Both problem sets and tutorial questions typically involve "traditional" organic chemistry questions: proposing mechanisms, predicting products, or suggesting synthetic routes to a target.

To students specializing in biochemistry, these typical organic questions about "flask" reactions or syntheses of uncontextualized targets often feel tangential or unrelated to the rest of their studies. Although biological examples are included in lectures, we thought there could be a benefit in asking students to consider and discuss biochemical questions from an organic chemistry perspective. Others have also discussed their own efforts to create an organic chemistry course tailored to biological scientists (5) with similar approaches to present the material as relevantly as possible. To be able to ask cutting edge and "interesting" questions, we felt these would need to be presented in a scaffolded way; in 2015, we began to consider workshops as a way forward. While tutorials would continue to focus on clarifying and consolidating the key concepts, the workshops would provide an opportunity to apply organic concepts in a more creative and open-ended way.

On-Site Workshops

As we began to design these workshops, there were several key aspects that we felt were critical:

- Fostering student collaboration and peer-to-peer learning
- Encouraging student-led discussion (with instructors in only a support or facilitator role)
- Focusing the questions on a piece of research in biochemistry or chemical biology
- Limiting the workshop to a self-contained session (i.e. there is no required preparatory or follow-up work).

The first two of these points relate to the inherent nature and strengths of workshops as an format that emphasizes student-interaction. The focus on research rather than biochemical topics more generally was made as a way to emphasize not only the relevance of organic chemistry to biochemistry, but the importance of it to modern research. Many students elect to specialise in STEM fields because of their interest in and enthusiasm for the fundamental questions these fields attempt to answer and the global challenges they can help to address. Introductory courses like the one here, in contrast, tend to focus on fundamental concepts and methods of problem solving, building the foundation for further study and deeper understanding. While the link between core concepts and big-picture research questions might be clear to instructors, they can feel disconnected to students which can be frustrating and demotivating as students begin study in higher education. By scaffolding from questions familiar to students to those answered by researchers in the department, learners appreciate the full scope of what their knowledge in first year already allows them to do. Because, as mentioned earlier, all students follow the same schedule of modules, the workshops also represent an opportunity for a synoptic view of the course: how physical chemistry experiments can provide insight to organic mechanisms, and both can shed light on biological processes. The final pillar of keeping the workshop self-contained was implemented as these sessions were initially added to the lectures and tutorials already planned. We wanted the workshops to enrich the student experience without a significant increase to student workload or stress. As the program of workshops expanded from one per year to multiple per term, these contact hours were re-allocated from lecture time rather than added on to the already present content.

While each workshop incorporates some specific elements depending on the particular topic at hand, they largely follow a similar logistical model in person. The cohort of students is divided into groups of 20–30 students. Each group has 1.5 hours scheduled to complete the workshop that has been designed to take approximately one hour. As they arrive, students are provided with a worksheet with the questions for discussion, and students choose where to sit. Tables are initially arranged to accommodate 5 students; however, students decide for themselves whom to work with, so it is not unusual for a table to be partially full or for students to push tables together to create a larger discussion.

The session begins with a brief introduction from one of the two instructors present, clarifying that nothing is required to be handed in at the end of the workshop; any written work or notes that students complete is for their own use and benefit. They are also reminded that the instructors are present to provide input or hints if they find themselves stuck. The students are then free to begin.

Each worksheet begins with questions very similar in style to questions that students will have seen on tutorial sheets or as example questions during lectures (e.g. identify the most acidic proton in a molecule or which of two species is the best nucleophile). As these examples also demonstrate, they typically involve a limited number of possible distinct answers. This facilitates the early discussion as each student can take a clear stance to explain and justify the other members of their group. From there, the questions become increasingly open-ended, encouraging peer-to-peer learning as they unravel more complex situations. For example, students are asked to suggest multiple potential mechanisms for a given enzymatic reaction and what experiments could allow them to distinguish between these hypotheses. This helps to break the misconception that mechanisms are something to be memorised but, rather, are intrinsically linked to the scientific method — something to be be hypothesized, tested, and observed (or not). Furthermore, this can serve as an opportunity for students to see the key concepts in action, consolidating understanding (6).

This format is grounded in Vygotsky's zone of proximal development (ZPD). Across the cohort, students are at various stages in their development from novice to expert organic chemists — this diversity is created not only by their individual school backgrounds and aptitudes, but also compounded by the variety of tutorial experiences depending on their grouping and tutor. By working in small groups in the workshop, students further along in the journey towards expert thinking can aid their colleagues (7), while the facilitators are present to become involved when these students reach their own ZPD.

The room is arranged such that there are paths for instructors to wander through and listen in on the discussions. Early on, students will often seek confirmation of answers to the right-or-wrong lead in questions. As the workshop proceeds, instructors will be invited to join in the discussions about which ideas are more or less feasible. Instructor contributions at these points are often questions or alternative ideas to further or redirect student discussion, rather than provide "answers." Throughout, groups may also send ambassadors to other groups to exchange ideas. At times groups also merge or bud into smaller ones depending on individuals' pace and communication styles.

Within this framework, four workshops have been designed to date that will be discussed briefly here. While contextualized in biochemical context, each focuses on one or more fundamental concepts of organic chemistry (8). They are spaced approximately evenly through the two terms of instruction: one in the middle of term and one at the end (sometimes taking place at the start of the following term).

Thinking in Three Dimensions

Just after the midpoint of the autumn term, this workshop focuses on concepts of molecular shape, isomerism, and stereochemistry. Centered on the use of Molymod molecular model kits, this workshop aims to clarify the links between three-dimensional structures and various two-dimensional representations (including zig-zag, Newman and Fischer projections). Among the workshops, this one is unique in not having a research focus in its current iteration. While there are biological molecules included, it was decided it was more important to focus on fundamentals, with the applications of these concepts being integrated into the following workshops.

DNA Alkylation

A capstone of the first term, this workshop aims to reinforce and refresh ideas including resonance, nucleophilicity, electrophilicity, and substitution. Questions begin with the concepts of three-dimensional structure covered in the previous workshop: challenging students to reconcile their knowledge that the DNA bases are flat in the experimentally-determined double helix structure with the fact that several atoms in the DNA bases have four areas of electron density (that they might predict to have a tetrahedral shape). This puts the concept of resonance (and its implications on structure and reactivity) at the forefront of their mind, scaffolding for the questions about DNA reactivity that follow. As students consider the ways that both biological processes and medicines make use of these tendencies, questions require balancing of inductive and mesomeric effects, using N-substituted nitrogen mustards (Figure 1) to foreshadow aromatic substitution reactions they will see in the second term. In the context of DNA methylation, students articulate why an S_N1 mechanism is an unlikely pathway for this process, while designing experiments that invoke kinetics or stereochemistry (using chiral methyl groups) that would support their answer.

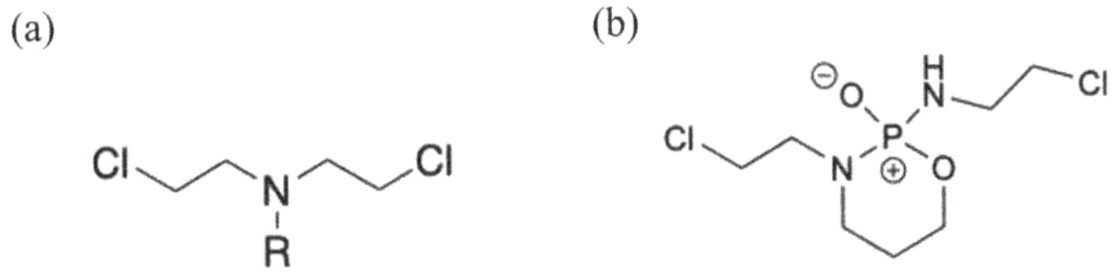

Figure 1. DNA Alkylation Agents. In the workshop focusing on the alkylation of DNA for biochemical and therapeutic purposes, one section features the nitrogen mustards. (a) The general structure of N-alkylated mustards that students are asked to rank in terms of nitrogen nucleophilicity, considering the inductive and mesomeric effects of the R group. (b) After considering simpler groups, students are asked to consider isofamide, a cancer chemotheraphy on the World Health Organization's List of Essential Medicines.

Glycosidases

The first workshop to be developed, this session focusing on the chemistry of sugar-processing enzymes takes place shortly after the midpoint of the second term, with an emphasis on alkene and carbonyl chemistry. The research it is based on comes from a group within the department where students are based (9), allowing for a great sense of vicarious discovery through the workshop and a glimpse of the research projects they can pursue in their final year. Again, this workshop opens with a reference to the previous workshop looking at how the inverting and retaining nature of glycosidases can provide clues to the enzyme-catalyzed mechansims by which the substitution proceeds. After receiving the structure of a molecule that was soaked into an enzyme crystal, students use PyMOL to investigate unexplained electron density in the enzyme active site (Figure 2) and propose a structure that accounts for the density as well as a mechanism for its formation. Finally, they evaluate molecules as potential inhibitors and substrates of the enzyme, identifying those that might be useful either as chemical biology tools to obtain crystal structures of the enzyme's catalytic cycle or as therapies for glycan-dependent diseases.

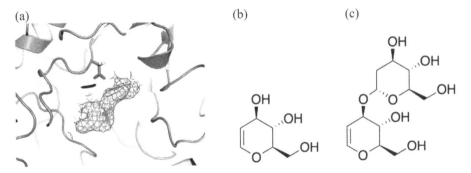

Figure 2. Sugars and Mimetics. (a) Students are presented with X-ray crystallographic data about ER α-glucosidase II (PDB:5HJO), including structures with small molecules bound. Students follow in the footsteps of researchers to explain why soaking the crystals with glucal (b) results in the density in of a larger shape—using chemical reasoning to propose a dimer (c).

Proteases and Esterases

The final workshop of the course once again centres on an enzymatic process, linking both to serine proteases that they have covered in other components of the course and extending the thinking to related enzymes such as those being developed to break down plastics (10). As a course retrospective, the starter question looks at the relative reactivity of various carbonyl functional groups and a standard base-catalysed mechanism for amide hydrolysis, before steadily introducing experimental data (*11*) to uncover the inconsistencies with their proposed mechanisms. Here the link between curly arrow mechansims of organic chemistry and the kinetic mechanisms of physical chemistry are emphasised in a more general sense, also linking back to the similar ideas in the DNA alkylation workshop.

The Pivot to Online

In spring of 2020 with the onset of the COVID-19 pandemic, as at universities around the world, it was necessary to redesign these workshops to be feasible in an online environment while maintaining or even enriching the benefits to students in this new format.

Philosophy of Pivot

Maintaining the content of the workshops was straightforward, although somewhat more challenging in the case of the Molymod building. In terms of format, we wanted to allow students to continue to synchronously collaborate in small groups and for the role of the tutor to remain that of facilitator.

Selection of Tools

To preserve these key features, we began to identify applications (*12*) that would enable two types of communication between the students to be adapted into an online format:

1. Language-based (either verbal or written)
2. Visual (i.e. for drawing and sharing structures)

For the first of these, we considered two options: Zoom and Microsoft Teams. While both programs (and other video-chatting applications) have quickly evolved throughout 2020 and beyond, at the time, breakout rooms were only available in Zoom. As this capability mirrors the students working in small groups around tables in the classroom (*13*, *14*), this was the deciding factor between the two applications.

With respect to visual communication, it was decided that a digital whiteboard would allow for students to display and interact with one another's ideas for chemical structures and mechanisms. Beyond this simply being possible, we wanted to be mindful of the diverse technology that students might be using to engage with the session, so that all participants could feel equally able and comfortable taking part. Zoom (like Teams) includes an integrated whiteboard. However, this whiteboard only lasts for the duration of the meeting and limits the drawing to electronic (with either a stylus or mouse). While a document can be shared and annotated, the annotations are not anchored to specific locations on the document limiting the Zoom-mediated collaboration on a document introduced via shared screen. Furthermore, the instructors can only see the whiteboard within the breakout room when they are in that breakout room.

Considering these limitations, we elected to proceed with a whiteboard external to Zoom: Miro (www.miro.com). In terms of accessibility, Miro is available as an application on smartphones and tablets, as well as through web browsers on a variety of devices. It allows the importation of documents and photographs, so students more comfortable writing with pen and paper can upload images of their ideas to the whiteboard rather than having to draw using the device. Because it is independent of the breakout room and Zoom meeting, instructors can have all active Miro boards open at the same time, regardless of which room they are in. The board remains available as a "living" artefact of the workshop for students to either review or continue collaborating asynchronously after the session is over.

The Miro board can be shared via screensharing within the breakout room in the same way as a document or native whiteboard can be, but it also has features that ease synchronous collaboration. "Show collaborators' cursors" shows the cursor location of everyone active on the board, enabling students to point out specific features as they discuss a mechanism or structure. In addition, each participant can "follow" someone else's view to see the exact zoom and portion of the board as the person they follow; conversely, it is possible to bring others to you, to see the view you currently have. These features, taken together, promote the activities of both creating and sharing visual ideas that we hoped to recreate from the on-site format.

Implementation

As in the case of the on-site workshops, students are scheduled to attend in groups of 20–30 with two instructors present to facilitate. To allow for potential delays from technical issues, each session is scheduled for 2 hours rather than 1.5, although the set of discussion questions was maintained at the same length. Before opening breakout rooms, an introduction is provided by one of the instructors, covering both the content-based material from on-site introductions as well as a description of the technical features of Zoom breakout rooms (such as the "Ask for Help" button). After students have an opportunity to ask questions, the breakout rooms are opened. Initially, we used random assignments to create rooms of roughly equal size (4–5 students). However, after Zoom introduced the feature that participants can choose their own breakout room, we trialled this feature to more closely reflect the on-site practice.

Once in the breakout rooms, the instructors circulate through the rooms to check again for questions and provide links to pre-prepared Miro boards. These are named to match each breakout room and have a copy of the workshop discussion questions already uploaded (a pdf of the questions is also available to the students via our learning management system). Afterwards, the instructors can return to the main room and observe all of the active Miro boards to identify groups that might be stuck or benefit from redirection. Because Zoom chat only allows communication within a room and the "Ask for Help" button notifies only the Host (but not any of the Co-Hosts), we have found it useful to have a "backchannel" open in Teams so that we can communicate about where support has been requested, regardless of our virtual location.

Lessons Learned and Ideas for the Future

Perhaps the greatest concern with the shift to online and hybrid workshops would be reduced engagement by the students. A similar proportion (>90%) of the cohort participated in the online iterations of the workshops compared to the corresponding on-site sessions. Data was not collected with respect to number of students contributing (as each breakout room was not observed at all times); however, based on the discussions with the students at the end of each session, students produced proposals and answers of a similar standard to those obtained in previous years in person. The average amount of time groups required, however, increased. Whereas an in person environment led to completion times of 1–1.5 hours, the extended time (included to accommodate technical difficulties) of 2 hours was typically required by the majority of groups (despite a paucity of problems with technology), with some requiring still further time.

In the first online workshop in spring of 2020, the cohort taking part had previously completed on-site workshops. Students (in anonymous feedback) and the instructors noted that discussions were slower to begin in the online format, with some students suggesting it was due to the randomly assigned groups, therefore not knowing one another or not having an already developed dynamic of collaboration. Although we purposefully designed the workshops to be self-contained, several students made the suggestion of providing a small activity for students to do individually beforehand to prepare them to participate from the beginning of the session.

At this first online workshop, there was also relatively little engagement with the Miro whiteboards, with students preferring to draw on paper and hold their work up to the camera to show one another. The few students who did make use of the Miro boards explained that they had previously used them in tutorials. Based on this feedback, we incorporated a description of the features and use of Miro to our introduction for the first workshop during the 2020–21 academic year.

A second situation that interfered with student discussion was the arrival of an instructor in the room, with any dialog stopping as soon as the presence was noticed. Whereas, in the on-site context, the instructor can unobtrusively circulate throughout the room and eavesdrop on student conversations, the instructor is decidedly present or not in the virtual breakout room. As a potential solution to this, we attempted only engaging in the breakout rooms when actively invited by the students inside. However, student feedback after this session indicated that students felt less well supported than when the instructors took a more proactive approach. One potential way to address this is to use a different video-conferencing software, such as Gather, that more closely mirrors the experiences of the classroom. Gather presents a video-game style where participants may move freely within a space, but only hear and see those they are close to. Instructors could more discreetly observe, and students are able to see the instructor approaching. We have recently trialled this technology for workshops with smaller groups (approximately 20 students). Students complemented the arcade graphics; in terms of practicalities, students and tutors found useful the wider communication available. It allows students to interact with other groups, reintroducing the possibility of ambassadors sharing ideas between groups, and groups evolving over the course of the workshop. All students could contact all tutors (rather than only the host as in Zoom), and it was possible for students to see where the tutor was, as in an on-site tutorial. A source of frustration for students (and instructors) in the breakout rooms in Zoom was that communication could only occur within a breakout room. So if students in room C asked for help while the tutors were actively discussing with students in rooms A and B, there was no way to notify the students in room C that someone would be there shortly. Although this potential delay in facilitator arrival was announced at the beginning of the session, it worked more smoothly in person and on Gather, where the tutor could send a chat to the other group to acknowledge the request for help.

Another factor that impacted student discussion during the workshop was the method of group assignment. Initially, only random assignment was available, and as noted above, students (who had previously done on-site workshops) preferred choosing their own group. When Zoom introduced the possibility for participants to select their breakout room, we enthusiastically implemented it, both in response to the student feedback and to mirror the in-person idea. Students organized themselves into groups of more varied sizes (from 3 –7), whereas groups were uniformly an intermediate size when assigned. Observationally, conversation amongst the groups started faster among these self-selected groups. This apparent advantage must be balanced with the benefits of working in new groups. Whereas tutorial groups often develop set social and power dynamics with leaders (*15*) and pairs (*16*), the workshops allows new combinations of individuals, providing diverse interactions and allowing students to assume different modes of participation and engagement. Interestingly, this cohort of students, who as first years in 2020–21 had fewer previous in-person interactions, said in feedback they would prefer being in random groups to get to know more of their coursemates.

Therefore, we think we will continue to allow self-selection in person, but alternate the formats when online.

Including the aforementioned introduction to Miro seemed to increase use across the groups (Figure 3). In fact some groups communicated almost exclusively via the whiteboard, only using the Zoom breakout room to interact with instructors. Students, as well as drawing directly on the whiteboard, incorporated their own drawings as well as figures they discovered online. It also allowed them to discuss multiple ideas simultaneously, with different participants focusing on different portions of the board at any given moment. As the connection was entirely separate, students whose Zoom disconnected were able to inform their colleagues through the more stable Miro link. The independence and longevity of the board was also capitalized on by the students, with participants returning to their shared ideas for weeks and months afterwards.

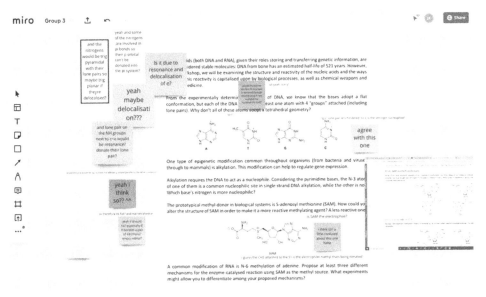

Figure 3. Student Miro Board from DNA Alkylation Workshop. Students share ideas via post-it notes on the board, as well as verbal discussion in the Zoom breakout room. They also paste relevant images or screenshots to share with their teammates.

As seen in Figure 3, students did research and incorporate online research during the workshop to inform their answers. In previous years, we provided references on the workshop problem sets for the research underlying the questions. When on-site, students did not look for the paper during the workshop; however, many groups immediately looked for the paper when doing the workshop online and then were not as open-ended in their ideas for questions. Based on this observation, during the later part of this year we removed the references from the problem sheets, providing them via our learning management system after the last group finishes the workshop. This change was effective; the first workshop after the change was the glucosidases which is the one most highly correlated with the associated publication, but students did not look for the paper or PDB files (even though they did seek out and incorporate other online resources).

Student feedback was similarly positive overall for both the on-site and online formats, and groups at other universities have seen similar success with synchronous online discussions (17). This spring a significant proportion of the students were able to complete the final workshop on-site. Their feedback indicated a strong preference for this format over online, stating that it was more "awkward" and "less engaging." The use of Miro as an additional way to communicate and create a shared artefact make it a useful tool that we are considering incorporating even in on-site workshops in the future. Particularly in the near future as even in-person experiences might require social distancing, this is a tool that allows students to visually collaborate whether 2 meters apart or in separate countries, enabling hybrid groups, too.

The most common student suggestion for improvement was for the production and release of an answer key, ideally with an exam-style grading scheme. Until now, we have not released model answers, preferring to provide feedback to specific student questions or answers. As several of the workshop questions are open-ended, we do not want the creativity of students to be limited to a single "right answer." However, we are considering implementing a more formal wrap-up, either at the end of the workshop or as a separate session to allow more time for independent consideration of the questions.

Students did not raise any issues with accessing the platforms described here (Zoom, Miro, and Gather); however some felt that the Miro whiteboard was "a bit unfortunate to navigate […] without a tablet." In terms of accessibility, a limitation of electronic whiteboards is the difficulty to produce screen reader friendly alt texts in real time. For this reason, their use in large classes should be carefully considered before implementation.

In a Likert-style question, more than 80% of students graded the workshops as either "very useful" or "a little useful;" when asked to rank lectures, tutorials, and workshops, the majority ranked them as the least useful of the three, with only two students identifying them as the most useful. In commenting on their rankings, students often acknowledged "workshops were fun, but I'm not entirely sure how relevant all their content was to prelim [exams]," as tasks were more open-ended. Both this qualitative and quantitative feedback are consistent with literature about active learning. Although it improves actual learning, students perceive it to be less effective than more passive learning methods such as lectures (*18*). This was appreciated by at least one student who commented in the feedback:

> *"I think, generally, it's good to create workshops that challenge people at a level above prelims, just so confidence is promoted about *actual* prelim exams."*

The research into this discrepancy between acutal and perceived learning recommends an intervention be implemented early in the course to describe active learning to students, addressing the gap between perceived and actual learning. While we have done this at the first workshop each year, we will remind students of this at every workshop in future years.

In all formats, the most common positive comments from students are that they enjoy working together and seeing the concepts they understand in the context of fundamental scientific questions and useful real-world applications. Recent literature has also reported that active learning helps to reduce awarding gaps for both low-income and underrepresented minority students (*19*), another worthwhile reason to further develop research-focused workshops in introductory courses in chemistry and more widely throughout STEM.

References

1. O'Dwyer, A.; Childs, P. E. Who says Organic Chemistry is Difficult? Exploring Perspectives and Perceptions. *Eurasia Journal of Mathematics, Science and Technology Education* **2017**, *13* (7), 3599–3620. https://doi.org/10.12973/eurasia.2017.00748a.
2. Kuhn, D. Thinking Together and Alone. *Educational Researcher* **2015**, *44* (1), 46–53. https://doi.org/10.3102/0013189X15569530.

3. Cormier, C.; Voisard, B. Flipped Classroom in Organic Chemistry Has Significant Effect on Students' Grades. *Frontiers in ICT* **2018**, *4*. https://doi.org/10.3389/fict.2017.00030.
4. Hein, S. M. Positive Impacts Using POGIL in Organic Chemistry. *J. Chem. Educ.* **2012**, *89* (7), 860–864. https://doi.org/10.1021/ed100217v.
5. Luh, T. Y.; Chou, C. K. Organic Chemistry for Biology-Oriented Students. *Trends in Chemistry* **2020**, *2* (3), 177–180. https://doi.org/10.1016/j.trechm.2019.10.003.
6. Reid, N. A scientific approach to the teaching of chemistry. What do we know about how students learn in the sciences, and how can we make our teaching match this to maximise performance? *Chemistry Education Research and Practice* **2008**, *9* (1), 51–59. https://doi.org/10.1039/B801297K.
7. Lou, Y.; Abrami, P.; Spence, J. Effects of Within-Class Grouping on Student Achievement: An Exploratory Model. *Journal of Educational Research* **2000**, *94* (1), 101–112. https://doi.org/10.1080/00220670009598748.
8. Duis, J. M. Organic chemistry educators' perspectives on fundamental concepts and misconceptions: An exploratory study. *Journal of Chemical Education* **2011**, *88* (3), 346–350. https://doi.org/10.1021/ed1007266.
9. Caputo, A. T.; Alonzi, D. S.; Marti, L.; Reca, I. B.; Kiappes, J. L.; Struwe, W. B.; Cross, A.; Basu, S.; Lowe, E. D.; Darlot, B.; Santino, A.; Roversi, P.; Zitzmann, N. Structures of mammalian ER α-glucosidase II capture the binding modes of broad-spectrum iminosugar antivirals. *Proceedings of the National Academy of Sciences of the United States of America* **2016**, *113* (32), E4630–E4638. https://doi.org/10.1073/pnas.1604463113.
10. Knott, B. C.; Erickson, E.; Allen, M. D.; Gado, J. E.; Graham, R.; Kearns, F. L.; Pardo, I.; Topuzlu, E.; Anderson, J. J.; Austin, H. P.; Dominick, G.; Johnson, C. W.; Rorrer, N. A.; Szostkiewicz, C. J.; Copié, V.; Payne, C. M.; Woodcock, H. L.; Donohoe, B. S.; Beckham, G. T.; McGeehan, J. E. Characterization and engineering of a two-enzyme system for plastics depolymerization. *Proceedings of the National Academy of Sciences of the United States of America* **2020**, *117* (41), 25476–25485. https://doi.org/10.1073/pnas.2006753117.
11. Zerner, B.; Bond, R. P. M.; Bender, M. L. Kinetic Evidence for the Formation of Acyl-Enzyme Intermediates in the a-Chymotrypsin-Catalyzed Hydrolyses of Specific Substrates. *Journal of the American Chemical Society* **1964**, *86* (18), 3674–3679. https://doi.org/10.1021/ja01072a016.
12. Kiappes, J. L. Using mobile phone applications to teach and learn organic chemistry. In *Technology-Enabled Blended Learning Experiences for Chemistry Education and Outreach*; Fung, F. M., Zimmermann, C., Eds.; Elsevier: 2021; in press.
13. Chandler, K. Using Breakout Rooms in Synchronous Online Tutorials. *Journal of Perspectives in Applied Academic Practice* **2016**, *4* (3), 16–23. https://doi.org/10.14297/jpaap.v4i3.216.
14. Saltz, J.; Heckman, R. Using Structured Pair Activities in a Distributed Online Breakout Room. *Online Learning* **2020**, *24* (1) http://dx.doi.org/10.24059/olj.v24i1.1632.
15. Bion, W. *Experiences in Groups and Other Papers*; Routledge: London, 1961.
16. Jaques, D.; Salmon, G. *Learning in Groups: A Handbook for Face-to-Face and Online Environments*; Routledge: Oxford, 2007.
17. Tan, H. R.; Chng, W. H.; Chonardo, C.; Ng, M. T. T.; Fung, F. M. Learning Online During the COVID-19 Pandemic: Using the Community of Inquiry (CoI) Framework to Support

Remote Teaching. *Journal of Chemical Education* **2020**, *97*, 2512–2518. https://doi.org/10.1021/acs.jchemed.0c00541.
18. Deslauriers, L.; McCarty, L. S.; Miller, K.; Callaghan, K.; Kestin, G. Measuring actual learning versus feeling of learning in response to being actively engaged in the classroom. *Proceedings of the National Academy of Sciences of the United States of America* **2019**, *116* (39), 19251–19257. https://doi.org/10.1073/pnas.1821936116.
19. Theobald, E. J.; Hill, M. J.; Tran, E.; Agrawal, S.; Arroyo, E. N.; Behling, S.; Freeman, S. Active learning narrows achievement gaps for underrepresented students in undergraduate science, technology, engineering, and math. *Proceedings of the National Academy of Sciences of the United States of America* **2020**, *117* (12), 6476–6483. https://doi.org/10.1073/pnas.1916903117.

Chapter 8

Teaching Chemistry Down Under in an "Upside Down" World: Lessons Learned and Stakeholder Perspectives

Elizabeth Yuriev,* Andrew J. Clulow, and Jennifer L. Short

Faculty of Pharmacy and Pharmaceutical Sciences, Monash University, Parkville, Victoria 3052, Australia

*Email: elizabeth.yuriev@monash.edu

Teaching first-year students presented additional challenges for online chemistry education in the times of COVID-19 pandemic. While addressing these challenges, we also had a year-long opportunity to try new teaching approaches, evaluate their effectiveness, adopt the things that worked and dispense with those that did not. Our students were able to provide us with ongoing feedback on their experience and had an ongoing and supported opportunity to get to grips with online learning. Together we – students, academics, and teaching associates – had an opportunity to grow as a community of teaching and learning. Having an extended two-semester-long and connected experience also allowed us to consider the pros and cons of our choices in three domains: technological (unfortunately this domain took priority at the beginning), pedagogical (this domain gradually reclaimed its primacy), and finally – affective. In this chapter we describe our decisions and experience in the context of these three domains. We also draw on perspectives of three groups of stakeholders: students, academics, and teaching associates.

Chemistry education has been making great strides in online spaces by taking advantage of technology (*1*) and developing online courses (*2*). Based on this accumulated body of knowledge and practices, the chemistry education community was able to react quickly to the onset of COVID-19 pandemic. Many practitioners shared their early experience of emergency remote teaching spanning a range of hot topics and hotly debated challenges: pedagogical course design decisions (*3*, *4*), delivery of labs online (*5*, *6*), student experience in general (*7–11*) and specifically student interactions online (*12*, *13*), assessment (*6*, *14*), and educators working together in a virtual international community of practice (*15*).

The sheer volume of what has been published in the first 12 months of the COVID-19 pandemic – in publicly-accessible and, increasingly, in academic space – makes the task of writing a chapter on online teaching and learning in chemistry extremely challenging. What can we say that hasn't already been said? What advice can we give or teaching implication can we highlight that would add

© 2021 American Chemical Society

to the body of knowledge that has been produced during 2020 around teaching and learning online? What is unique and informative about our own experience? One aspect of our experience is that we taught first year students. While not unique, it presented additional challenges of supporting a student population lacking prior higher education exposure and the associated expectations, both of the university and of themselves. Another aspect is that the start of the pandemic coincided with the start of the academic year in Australia, almost to the day. And even though we did not know it at the time, we had a whole academic year ahead of us to teach fully online since Melbourne, alone among Australian cities, had two major COVID-19 waves during 2020, one at the onset of each semester. Lastly, the Monash University Bachelor of Pharmaceutical Sciences (BPS) degree program (*16*) entails two consecutive general chemistry units in the first year (Monash "unit" is an equivalent of "course" in the United States). These units are taken by the same cohort of students, convened by the same academic, and the workshop sessions are facilitated by the same group of teaching associates. The confluence of these three aspects of our's and our students' experiences meant that we had an opportunity to try new methods, evaluate their effectiveness, adopt the things that worked and dispense with those that did not. Our students were able to provide us with ongoing feedback on what worked and what did not work for them. They also had an ongoing and supported opportunity to get to grips with online learning. Together we had an opportunity to grow as a community of teaching and learning. Finally, having such a longitudinal and connected experience allowed us to consider the pros and cons of our choices in three domains: technological (unfortunately this domain took priority at the beginning), pedagogical (this domain gradually reclaimed its primacy), and finally – affective. In this chapter we describe our decisions and experience in the context of these three domains. We also draw on perspectives of three groups of stakeholders: students, academics, and teaching associates.

Context and Data Collection

Teaching Philosophy and Approaches

Teaching within the Bachelor of Pharmaceutical Sciences (BPS), Faculty of Pharmacy and Pharmaceutical Sciences, Monash University, has undergone two major renewal actions in recent years that prepared us to a great degree to handle the pivot to online teaching and learning. In 2012-2013, a whole-of-institution teaching overhaul resulted in significant flipping of the curriculum (*17–20*). The teaching and learning model that has been adopted by the Faculty is called DEAR (*21*). The learning cycle that students undertake for each topic includes the *Discovery* stage. Then they *Explore* new knowledge in interactive lectures and *Apply* it in small classes (e.g., workshops). Finally, they *Reflect* and *Revise*, while preparing for in-semester and end-of-semester assessments as well as engaging in skills coaching and formal reflective practice through e-Portfolios.

In 2018, a newly re-structured curriculum was rolled out, with a focus on professional and employability skills: communication and teamwork, problem solving and critical thinking, and laboratory skills. In addition to the undergraduate programs, we have developed and have delivered multiple iterations of a massive open online course (MOOC) "The Science of Medicines" (*22, 23*). All of these changes and activities involved an extensive development of online resources: videos, interactive quizzes, wikis, crosswords, Moodle Lessons, to name just a few.

As a result, we were arguably in a relatively advantageous position to tackle the pivot to emergency online teaching. However, together with the rest of the higher education field worldwide, what we were not as well prepared for was the speed with which the changes had to be made.

Furthermore, our arsenal of resources was developed for asynchronous delivery. The need to do some of the teaching and learning synchronously still left us faced with many challenges.

Units of Study

This chapter describes teaching and learning experience in first year general chemistry units with a focus on physical chemistry. Both are *core* program units, which means that they are prerequisite for students to progress through the degree program. Historically (pre-COVID), these were flipped units with traditional teaching activities such as interactive lectures, tutorials, workshops, and laboratory classes. In these units, the students are assessed on both theory, laboratory practice, and graduate skills.

Data Collection: Learning Analytics

To collect the data on student engagement with teaching resources on the Learning Management System (Moodle) and the video hosting platform (Panopto (24)), we have used the learning analytics tools available on both platforms. Learning analytics data collection for quality assurance and education research purposes is approved by the Monash University Human Research Ethics Committee (MUHREC, project approval 16974). Students are informed about the project during year one orientation and are provided with an opt-out link via a PharmSciHub Moodle site. They can withdraw their data from analysis at any stage during their undergraduate studies and afterwards, while they still have access to the Moodle platform.

Data Collection: Stakeholder Voices

Two instruments were specifically developed to get an insight into the perceptions of teaching and learning online during the pandemic-affected semesters. Teaching Associates (TAs), involved in teaching both semester 1 and semester 2 chemistry units were interviewed after each semester. Students were invited, at the end of the academic year, to complete a Qualtrics survey about their experiences of learning online during 2020. Both projects were approved by the Monash University Human Research Ethics Committee (project approvals 26828 and 23346, respectively).

Dilemmas, Decisions, and Teaching Implications

This section consists of three parts. We will start by briefly describing technological and pedagogical decisions we made, and sometimes the dilemmas we faced when making them. This first part will provide the academic stakeholder view on how we have re-imagined the usual elements of the unit – discovery, interactive lectures, workshops, laboratories, and assessment – in the online space. Then, we will present findings from the TA interviews and student surveys to give insight into the other two stakeholder perspectives on the online teaching and learning experience.

Teaching Online

Discovery

The discovery stage was relatively easy to adapt to online modality as it was designed to be delivered fully online and asynchronously in the first place. The discovery stage usually includes "something to watch", "something to read", and "something to do". For something to watch, we have previously created a range of Animations (Figure 1). These came in handy during 2020, when our ability to draw things in front of students, in order to explain complex concepts, was terribly diminished.

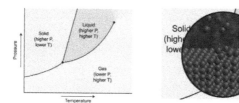

Figure 1. Animations used for Discovery. Snapshots from the animation of a single-component phase diagram animation are shown as an example.

For something to read, we use traditional textbooks. During lockdown restrictions, students could not come on campus to access the library. So, for reading, we looked for quality online resources to give students short bursts of reading material. The LibreTexts website (25) provided most of the required reading. And finally, for something to do, there was usually a short check-your-learning quiz.

Content Delivery, or "Interactive Lectures"

The first of the two biggest dilemmas came up when considering the interactive lectures (Figure 2). Should we run them synchronously or asynchronously? If they are pre-recorded for asynchronous delivery, should they be scheduled or not? This was likely one of the biggest dilemmas of 2020 for most educators internationally, see for example Ref. (4). For synchronous delivery, we briefly considered running live lectures, streaming them, and recording for students who could not watch live. Recording lectures was not a new practice in Monash in 2020, and our past experience showed limited student attendance of live lectures and their preference for having access to recordings (20). Based on that experience, we opted for pre-recording lectures, cutting them into short single-concept videos, and integrating them with self-testing activities (Figure 3). These pre-recoded lectures were uploaded to Panopto (24), the video hosting platform used by our university. The greatest advantage of pre-recording the lectures was that they allowed students to work at their own pace and at the time of their choosing. Considering the multitude of personal issues induced by the pandemic and barriers experienced by students while studying online (*vide infra*) as well as a wide range of students' personal preferences, the decision to use pre-recorded lectures was sensible and balanced. Various, and sometimes conflicting, student preferences were still manifested in the end of year survey, for example:

"The discoveries and lectures are all online, which makes uni life easier"

"[I prefer] live (rather than recorded) lectures (for all units)"

Figure 2. Possible delivery modes considered for interactive lectures.

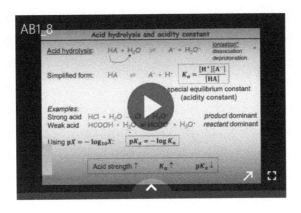

Select the strongest of the following weak acids

Select one:
- a. Phosphoric (pK_a = 2.1)
- b. Carbonic (pK_a = 6.4)
- c. Benzoic (pK_a = 4.19)
- d. Nitrous (pK_a = 3.37)

Check

Figure 3. Snapshot of interactive lecture components. Top: a pre-recorded lecture segment; Bottom: a corresponding test-yourself question.

The duration of the recordings in each lecture was about 30-40 minutes, leaving time for students to solve problems and answer quiz questions. Students had unlimited attempts at the quiz questions, which contributed a very small mark to their final unit grade. A very important element of these lectures was a request for students to note the 'muddiest point', that which they found most difficult to grasp. These questions were then raised and addressed at a live Q&A session the following week. This feature was later noted by students as one of the highlights of the chemistry units and one of the strongest contributors to student learning.

There are two additional aspects associated with using pre-recorded lecture videos. Firstly, should the videos from the previous year (in this case, those recorded in 2019) be used? Some of our colleagues selected this option. It is our opinion that such an approach is not appropriate. Students studying online in 2020 were already disappointed by not being able to learn on campus, meet new people and make friends. They were already experiencing stress and mental health issues (as we will demonstrate later). The very least they should receive is a tailor-made instruction designed with their needs and circumstances in mind.

Another aspect is related to an organizational, or motivational – depending on how you look at it – question to ponder. This is: should these pre-recorded lectures be scheduled or not? Largely this decision was out of our control, because the University and to some extent the Faculty made the decision about timetabling. In the first semester, when any decision had to be made very quickly, and the timetable already existed, these lectures were left in the timetable. Learning analytics showed that most students completed or at least started these interactive lecture activities towards the start of a week as per their timetable (Figure 4).

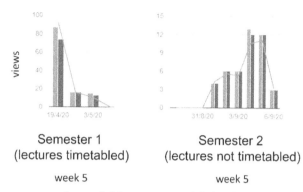

Figure 4. Student engagement with recorded lectures, exemplified with week 5 learning analytics. For each week: left column, views and downloads; right column, unique viewers.

In the second semester, when the University and the timetabling administrators had more time to plan for pandemic/lock-down conditions, the decision was made not to timetable these activities. On one hand, logistically it made sense. Putting together a timetable was easier and allowed for more flexible and streamlined scheduling. It also gave students more flexibility with respect to their time management. However, learning analytics are already showing us that, while some students actually did complete these activities earlier in the week, a significant proportion of students left it until the very last moment (Figure 4). Some students noted the lack of scheduled lectures and expressed the preference for scheduling in their end-of-year survey, for example:

"... a bit stressful, mostly cause I procrastinate quite a fair bit since most lectures are recorded and there's no fixed deadline for lectures."

Q: What changes would have improved your online learning experience?
A: *A more rigorous timetable, set times assigned to watch lectures*

These data show that if, in the future, lectures are to be online and are not to be timetabled, proactive measures are needed to make sure that students, particularly transitioning first-year students, are staying on top of their learning.

Small-Group Activities (Workshops)

The biggest smorgasbord of choices, and therefore decisions to be had, were around small class activities. The decisions, largely because of the speed with which they had to be made, revolved around technology choices: should it be Zoom? or Teams? Or should we use a forum? But once the technological decisions were made, the personal orientations, of both instructors and students, came into play. And here we come to the second major controversy of teaching and learning online in 2020: should the webcams be ON or OFF? And the answer to that particular dilemma in turn led to pedagogical decisions. How do we ensure effective and fair teamwork? Do we assess participation? And if we do, then how do we do that in the digital environment?

When it comes to cameras on or off, there are strong opinions either way (26). There are good reasons to allow students to keep their cameras off in large classes. However, for small groups, our experience based on both students' and TA's perspectives (discussed below), is that the use of cameras is a significant factor in maintaining effective teamwork, better interactivity and social climate in classes, all of which contribute to improving learning.

The other tool that allowed students to work effectively in small-group activities was the use of Google Docs in parallel with Zoom. Google Docs allowed students to interact verbally as well as digitally. Importantly, each Google Doc performed several additional functions. It served as a permanent record of the work students had done in class, which they could return to later during study or revision. It allowed TAs to monitor the work of several small groups, while only being able to be present in one Zoom break-out room at a time. It also provided academics with the record of student contributions to the class work, thus enabling us to identify and reach-out to potential students-at-risk.

Laboratories

For chemistry educators, laboratories were probably the most difficult aspect of moving teaching and learning online. In our units, we opted for recording the videos of the experiments (Figure 5), providing students with data to process, and supporting their calculations and conceptual questions related to labs with SCORM exercises (Figure 6). Videos allowed us to show students what happens in an experiment. Regrettably, this approach suffers from not being able to support students in developing their practical skills.

Figure 5. A front-on screen shot of one of the video demonstrations.

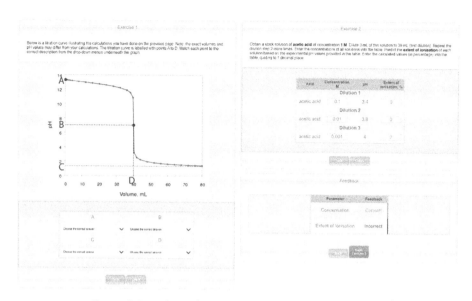

Figure 6. Snapshots of a SCORM exercise on acids and bases.

SCORM (Sharable Content Object Reference Model) is a programming language that can be used to develop activities for students to perform data processing with randomly generated data sets. SCORM applications are great for practice, in that students can look at the same experiment multiple times and be provided with new data sets. SCORMs automatically generate feedback, thus reducing the need for TAs to tell students whether they are on the right track. It is possible to incorporate small quizzes into SCORMs that go beyond just calculations. Finally, SCORMs integrate very well with most Learning Management Systems.

Communications

Most of these activities – interactive lectures, workshops, online labs – happened every week. And so it was very important to help students stay organized. With that in mind, we provided students with weekly activity tables (Figure 7). This was done similarly across all undergraduate units and thus created expectations on the part of the students and consistency for them. In addition to the activity tables, we also provided students with weekly wrap-up and planning recordings.

Figure 7. An example of a weekly activity table.

Assessment

Assessments, particularly the end-of-semester exams, are a big issue that deserves a chapter (or a book?) of their own. They presented us with a multitude of questions and choices to be made: Do we need invigilation or not? Our Faculty went for an un-invigilated option. Is the exam going to be open book? An un-invigilated exam is an open-book exam by its very nature. Should the exam be open note, considering that it is already open book? We encouraged students to prepare summary notes as they would do any other year for an open-note exam. The feedback received from students was that preparation of notes was extremely useful, even in an un-invigilated open-book assessment. Next came designing the exam itself. Do we randomize questions? Do we set time limits on each question? Do we allow free navigation, where students can go up and down the exam? Or restrict them to answering questions sequentially, so that students would only be able to go forward? And finally, how do we ensure academic integrity?

In the end, the experience of administering un-invigilated open-book exams left us with more questions than answers. At the time of writing, we are about to embark on a second year of online teaching and learning, with most of the activities still delivered remotely, and the laboratory classes being the only ones planned for on-campus delivery. The process of designing an appropriate examination regime is still currently underway, but one lesson from the previous year is clear: the most important aspects of un-invigilated, open-book online assessments are that they have to be robust, fair and not overwhelming to students.

Feedback from Students

After Semester 1 was finished, we asked the students to answer three questions about the unit: What to keep? What to start? And what to stop? Thirty-six year-one students provided responses. Student answers were largely consistent among them, and the common representative themes are summarised in Table 1. Interestingly, for the 'keep' part, the answers were largely centred around technology and organization. However, for 'starting' and 'stopping', that is what student felt were missing or things that they did not like, the suggestions were very similar to what we get in a so-called normal year. Students want: more model answers, more past exam questions. Students do not particularly like being marked on their problem-solving process and prefer being marked on whether their answer is correct or not. And, as in any so-called normal year, we will continue the conversation with students about why we do certain things in the way we do. However, this 'normalcy' of student responses indicated to us that they were able, to some extent, to get to grips with the disruption of fully online study and managed to focus on the tasks in front of them

Table 1. Student and TA Responses to the Keep-Start-Stop (27) Questions

	Common student responses (N = 36)	*Common TA responses (N = 9)*
Keep	SCORMs Zoom & Google Doc Integrated recordings and quiz questions Weekly activity tables	Zoom for small classes Communication back-channels
Start	Model answers for quiz questions More past exam questions More conceptual questions	Technology solutions for seeing student thought process Cameras ON in small group work to see whether students understood
Stop	Forums as communication channel for workshops Marking the problem-solving process Uploads of workings to the forum	Non-engaging students affecting the experience for other students Students not working with each other but working concurrently on the same tasks

At the end of Semester 2, a more detailed survey was undertaken probing student experiences and perceptions of a whole academic year spent entirely online. Fifty-seven year one students completed the survey. The detailed analysis of student responses is currently being performed and will be published elsewhere. Here, we focus on student perceptions about the activities that contributed most to their learning and what factors significantly reduced their ability to study (Figure 8). The answers to the first question are critical to our design of the student learning experience going forward. The answers to the second question, combined with qualitative data (Table 2), give us a deeper insight into student overall experience.

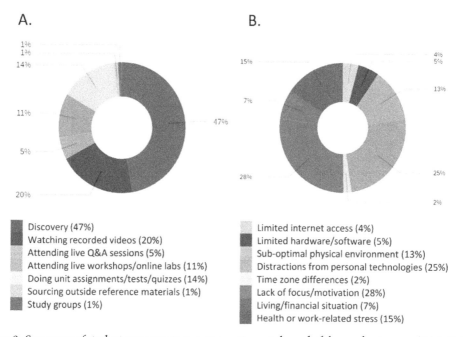

Figure 8. Summary of student responses to survey questions at the end of the academic year. (A) Activities that contributed most to online learning. (B) Factors that significantly reduced the ability to study.

Table 2. Common Themes in Student Responses to Open-ended Survey Questions

Issues	Student quotes
Isolation and lack of human interaction	Cannot make new friends Feeling disconnected from lecturers and students Human interaction makes the learning experience 100% better so without it, felt very pointless and dry.
Lack of privacy	To have a less distracting learning environment. I live with my family and I have 4 sisters, so it can get noisy at times and we are all sharing the internet bandwidth.
Merging of study and personal spaces	I think it's easier to stay on top of your work if where you work is separated from where you eat, sleep and relax. There was no separation between uni and the rest of my life. Uni work felt like I was bludging at home, and relaxing felt like I was at uni and had to go do more work. It all just felt the same

Table 2. (Continued). Common Themes in Student Responses to Open-ended Survey Questions

Issues	Student quotes
Workload	*The sheer workload has been unforgivingly overwhelming this semester*
Stress and mental health issues	*Mental health issues - being cooped in home all day whilst having to just study all day took its toll* *I feel overwhelmed* *since I am living alone, the lockdown situation and online learning has been overwhelming and so it affects my mental health which affects my motivation to study*
Concerns about lack of lab skills	*… lots of labs have been canceled. And this makes me very hard to understand the lab skills* *Not being able to apply theoretical concepts in laboratories or other interactive opportunities, which sometimes makes it challenging to understand certain concepts.* *On campus Lab, pleeeeeeeeeeeeeeeeeeeeease.*
Frustration about non-participating team members	*I feel like online learning has given more opportunities for people to avoid their responsibilities in group based work/assignments and unless I get to choose who I work with, chances are I would be the one that has to put more effort than others into completing the task satisfactorily. I believe that online teaching has certainly given more "cover" for students to act selfishly which otherwise wouldn't happen to the same effect in a physical environment.* *Working in teams with strangers we have never met and are not willing to contribute has been the most stressful and annoying factor* *Dealing with teammates in workshops who refuse to contribute*
Cameras ON/OFF	*[need] Students having their videos on and engaging in conversation more* *a bit hard when I was in classes with people who I didn't know and didn't even know what many of them looked like.* *It has been very frustrating to deal with quiet breakout rooms*
Motivation	*Initially, I could cope with online classes as I was enthusiastic about starting uni and determined to make a good start. However, as the year progressed, I found it harder to keep myself in check and found myself slacking more. With face-to-face classes, I find it easier to check in with my peers on their progress and to keep myself motivated.* *I am generally more motivated in face-to-face learning because it is much easier to work with your peers in person and there are generally no other distractions such as those that there may be at home. For online learning I find that I tend to lose focus and tune out much more easily therefore decreasing my motivation.* *In-person I feel there is more competition as motivation for me to perform better, seeing my classmates and wanting to do better. Online, I don't know or feel like I have classmates so I feel very alone in studying so I have very little motivation. There's also feels like there's no physical goal for me to look up to, like a teacher.*

Table 2. (Continued). Common Themes in Student Responses to Open-ended Survey Questions

Issues	Student quotes
Positive experiences	*My online learning experience was mostly positive as I had adequate access to materials and didn't feel it was compromised majorly aside from no physical presence in classes/lectures.*
	Zoom classes have really improved my confidence, because I know that I never would've contributed to open discussions or asked questions in a lecture theatre but I have no problem doing it over Zoom
	I feel generally good as I can avoid commuting and get more rest. I also find it very draining to be around people, so online learning is less stressful than on campus learning
	It's hard but rewarding
	Overall, quite satisfied. I made sure to stay on top of my work and engage during classes.
	As for online learning, since I am a shy girl, I am afraid to share my opinion face-to-face, however, I am confident to talk something using chat. And also, as for online learning, I can watch lecture video as many times as possible at home to ensure I get the knowledge
	Online learning should not disappear with the return to campus, I think it gives a lot of students the option to learn flexibly, depending on their own learning habits.

Feedback from Teaching Associates

Similar to students, TAs were asked to provide a brief Keep-Start-Stop feedback at the end of the Semester 1 (Table 1) and a more detailed in-depth debrief at the end of Semester 2 (Table 3). Similarly to the students, only a small section of feedback had to do with technology concerns. What worked for TAs was having Zoom for small classes where they could catch mistakes and provide immediate feedback. They appreciated regular and timely communication, in particular communication back-channels between TAs and between TAs and the lecturers during workshops. When it came to the 'start' part of their response, it became clear that teaching online for less experienced educators is a huge challenge, even if they sometimes do not recognize it themselves. Seeing students thought process and facilitating problem solving is not easy, even in a face-to-face classroom. The online environment makes it manyfold more challenging, particularly when cameras are off. How do you judge whether students understood your explanation? That question was absolutely on the top of TAs' list of questions. In the 'stop' part of their response, the lack of teaching experience was again clear. What TAs wanted to stop was students being students: saying 'Yes' just to get out of class and not working with each other. This is no different in an online classroom compared to face-to-face, just that much harder to communicate directly in some cases online.

Table 3. Common Themes in TA Responses to Interview Questions

Issues	TA quotes
Developing pedagogical content knowledge	*… after they were explaining their solution and then I would reiterate the point that they did really well, just to revoice and reinforce what they did well. And then in terms of expanding if they had a good skeleton, but they were missing extra things, I would then expand on their solution and say "Great explanation, I'd add this, this and this to get the full … explanation."* [Katherine]

Table 3. (Continued). Common Themes in TA Responses to Interview Questions

Issues	TA quotes
Engaging with students' non-verbal ques	*And for Zoom I think that very good thing is I can see their facial expression. Whatever I try to rephrase something I always look at their facial expression.* [Cate]
Establishing an emotional connection with students	*That's what I… that's what I like to do in terms of that emotional component… When I first started, when I started my workshop, so I'd always ask them how are they, how they are doing. And I obviously.. you know in Zoom it's not… they're not like going to start therapy session, but they say, you know, 'good', or they say 'they're tired' or something like that. And then I would say… it was like I'd kind of like … 'Yeah, me too', like you know. 'It's… it's it difficult, so I understand where you're coming from'. Just to emphasize that I understand them and then unlocks it coming into the workshop expecting them to perform like you know at 110%. It's more like 'I understand how you feel, let's just work together and help each other.'* [Katherine]
	… the talkative group were constantly asking me about you know, even things outside. Like sometimes they asked me what was my experience of studying face-to-face when I was an undergrad. So last conversations that were not necessarily related to the workshop, but it did help me connect with the students and in terms of providing feedback and motivating them. Just the fact that I was constantly going between… between the breakout rooms. Like they knew that I was there. So it made it easier for them to ask for help. Even the shy students that didn't necessarily feel OK to talk. I would just say 'I know this is a tricky question. But looks like you've made a good attempt at the first part. How are you going?' … I felt that if I did start with uh encouraging compliments and say their name, even if they're shy student that have webcam off… [Mary]
Facilitating students establishing connections with each other	*As I got to understand the students more in terms of their personality, like level of communication and academic performance, I would actually couple students that were… you know… more interactive online with those that were more quiet. And so that would facilitate communication … I think students communicate quite readily, when they communicate with each other. … So I would take into consideration the personality and the level of academic ability and I found that really helped because a lot of students… I mean, I mean, just in general, if you know something, you want to communicate your knowledge. … I found that to be very beneficial for students. They got out of their shell and then, once other students that struggled understood it more, they actually even started to encourage discussion. I found that towards the end of the semester, when I even started to do that and I even I noted, when they did that actually… came into the Zoom session like great assist, really good teamwork, things like that, reinforce that positive discussion that students were encouraged to do that…. I actually thought where the struggling students at the beginning, really contributed and really helped other students towards the end. Because it was just like teamwork. I think they felt like they were 'one', like a team as opposed to just individuals. So that's how I encourage discussion in that way.*
	I think what I achieved well is creating that… that space for students to feel comfortable with learning and asking… asking questions and making mistakes. [Katherine]

Table 3. (Continued). Common Themes in TA Responses to Interview Questions

Issues	TA quotes
Differentiating approaches	*Yeah, with the talkative group, ... I did feel like they were involved. Even though each person is assigned one particular problem, they were involved with other problems enough that they... But then I would say they are also strong group of people, so that's why it was a bit easier, whereas with my other group, I had a few strong members and a few weak members. But overall, they were very quiet, so if they were asking each other questions, it was because I was forcing them to ask each other questions and their questions were usually really simplistic ... [Mary]*
	There was like a couple of students who are really engaged, I guess. And they were willing, always willing to kind of answer or even sometimes ask questions themselves but... the students, which you probably picked the first couple of weeks, were always very hard to get... They... they were expecting me to ask them to ask a question. I still found that they weren't necessarily proactive with asking the questions so much. It was still... I was sitting in the room going 'alright, so I guess, Miriam, you are asking a question on this one. What... do you have any questions?' [Colin]
Cameras ON/OFF	*one of the biggest difficulties was missing non-verbal communication to students due to lack of camera usage compliance throughout the year, but especially in Zoom Sessions. This made it very hard to gauge how well students were responding to content or guidance; whether there may be having difficulties with what is being discussed, or whether they might be genuinely tired or frustrated, and thus easing off may be a better approach. As a TA, this meant that I often felt like I was 'shooting in the dark', as well as uncertainty as to what I was doing (or not doing) was effective. [Billy]*
	Again, my quiet group most of the time... none of them had their camera on. My talkative group did have their camera on again. I think it's just a personality thing, but as a result of that my quieter group, I don't have much of, I guess, a personal connection with them or any conversation other than the work. [Mary]
	It was hard. It was really, really hard. I think particularly in my groups no one wanted to use their cameras. And there was probably... In each of the groups, there was one or two people who ... Don't know if extroverted is the right word... But they were more open and kind of willing to answer but a lot of the times you kind of felt like you were speaking into the void a little bit. [Interviewer: So how did you manage it?] Mainly it was calling on people by name. So it was very much like 'Well? Ok, I'm going to volunteer, say Jimmy. If you want to, can you answer this question for me. And then, yeah, there would still be a pause before they would be like alright. [Colin]

The detailed thematic analysis of TA interviews will be published separately. Here we present common themes in these interviews that complement the picture provided by student survey responses and our own observations (Table 3). More in-depth interview answers revealed that TAs were developing approaches to reach students online where non-verbal cues were limited, particularly the body language and particularly when students elected not to turn on their webcams. TAs talked at length about working on establishing an emotional connection with students as well as facilitating students establishing connections with each other.

Stakeholder Perspectives

There are three players in this game we call higher education: us (called faculty or academics, depending on geography), students, and our sessional staff (teaching associates).

Students

Students have shown to be largely resilient, adaptable, and understanding during the year of online learning. But the transition to fully online learning was hard on many of them, particularly on first-year students who had at most one day of on-campus experience. They learned to grapple with, and even enjoy, accessing classes, their instructors, and support services online. However, they also missed the social – and educational – aspects of face-to-face classroom experience.

It was not a revelation, but we have definitely confirmed that structure is very important. That involves clearly sign-posting all activities and assessments, informing students about what happens when, keeping it simple, keeping it up-to-date, keeping it all in one place. It was also very important to not use too many different technology tools, at least not all at the same time. Communication is very important, it has to be a regular two-way communication between students and instructors. It has to support student-to-student communication, both formal and informal. Communications in the learning management system should include instructions for students on 'where, when, what, and how'. The communication must also include early and regular feedback *from* students on what does and does not work. Communication could be recorded, but the live sessions make students feel teachers' presence. And finally, student well-being is critical and therefore pastoral care is critical.

Teaching Associates

Teaching associates have clearly risen to the challenge. They have mastered the technology and tackled the challenges of encumbered online communication. They have developed approaches to engage with students in an environment that was unfamiliar to both students and themselves. TAs have not only identified areas of further improvement in course design, but also the gaps in their own preparedness to teach online.

Teaching associates are an absolute cornerstone of our workforce, when it comes to teaching both on campus and online. So it is critical that we support them both before, during, and after each class, particularly for online classes. They need special training for online teaching that goes beyond how to use Zoom. They need training in appropriate teacher discursive moves and facilitation techniques.

Academics

We learned that we must adapt our pedagogy, and not just re-use our old PowerPoint lecture files. We must look for resources as we cannot develop everything ourselves. We should share our resources, as good citizens of the community of practice. We also could take this great opportunity to experiment with technology, but not go crazy with it. We need to continue engaging with education designers. We need to be flexible and agile. We need to be organized. Importantly, we need to remember to look after ourselves.

And finally, a big question: did students learn in 2020? Based on the experience in the two chemistry units discussed in this chapter, the answer is Yes. But, it is possible that some students achieved better learning outcomes than they would otherwise, whilst others were not as successful as they would have been in a face-to-face on-campus experience. So does it mean that the online environment favors students with greater prior learning and stronger self-regulation? Possibly. And here we may have a wicked problem that we must be aware of going forward.

Summary

Teaching and learning in the COVID-19 pandemic was a valuable lesson in what does and what does not work well in the online teaching environment. This natural experiment has provided a proof of concept that online teaching and learning is effective for some elements of university courses, even those courses that require students to develop practical skills, such as laboratory skills. We found that discussion forums did not work well when used synchronously and that pre-recorded lectures did not work well if they were not timetabled. On the other hand, Q&A Zoom sessions and weekly structures (activity tables and wrap-up videos) worked extremely well to keep students engaged, informed, and connected. Post-pandemic, keeping online what worked online and returning to the physical classroom what did not work online will allow instructors to truly take forward the concept of blended learning (*28*).

Curiously, and maybe unexpectedly, the urgent pivot to remote teaching caused many academics, especially those outside of the teaching-only or education-focused throng, to return to the values of pedagogy. All of a sudden, many had to re-examine "how we teach", not just "what we teach" (*29*). We note that this return to valuing 'good old' quality teaching may be another lesson that had to be learnt. Technology, format, and modality are important, but teaching quality must reign supreme.

Acknowledgments

We would like to acknowledge the degree program team, the education design team, and the Associate Dean Education Prof. Paul White for all their effort and support. EY would like to thank her amazing team of teaching associates Kimberly Vo, Caely (Shuqi) Chen, Matt Burton, Matthew Challis, Matthew Urquhart, Tom Eason, Mahta Mansouri. Great thanks to Drs. Trayder Thomas and Tamir Dingjan for developing animations, SCORM exercises, and laboratory videos. This work was partly supported by the Australian Research Council through Discovery Early Career Research Award awarded to AJC (DE190100531).

References

1. Christiansen, M. A., Weber, J. M., Eds. *Teaching and the Internet: The Application of Web Apps, Networking, and Online Tech for Chemistry Education*; American Chemical Society: 2017.
2. Sörensen, P. M., Canelas, D. A., Eds. *Online Approaches to Chemical Education*; American Chemical Society: 2017.
3. Rupnow, R. L.; LaDue, N. D.; James, N. M.; Bergan-Roller, H. E. A Perturbed System: How Tenured Faculty Responded to the COVID-19 Shift to Remote Instruction. *J. Chem. Educ.* **2020**, *97*, 2397–2407.
4. Kyne, S. H.; Thompson, C. D. The COVID Cohort: Student Transition to University in the Face of a Global Pandemic. *J. Chem. Educ.* **2020**, *97*, 3381–3385.
5. Wild, D. A.; Yeung, A.; Loedolff, M.; Spagnoli, D. Lessons Learned by Converting a First-Year Physical Chemistry Unit into an Online Course in 2 Weeks. *J. Chem. Educ.* **2020**, *97*, 2389–2392.
6. Harwood, C. J.; Meyer, J.; Towns, M. H. Assessing Student Learning in a Rapidly Changing Environment: Laboratories and Exams. *J. Chem. Educ.* **2020**, *97*, 3110–3113.

7. Simon, L. E.; Genova, L. E.; Kloepper, M. L. O.; Kloepper, K. D. Learning Postdisruption: Lessons from Students in a Fully Online Nonmajors Laboratory Course. *J. Chem. Educ.* **2020**, 97, 2430–2438.

8. Rodríguez-Rodríguez, E.; Sánchez-Paniagua, M.; Sanz-Landaluze, J.; Moreno-Guzmán, M. Analytical Chemistry Teaching Adaptation in the COVID-19 Period: Experiences and Students' Opinion. *J. Chem. Educ.* **2020**, 97, 2556–2564.

9. Ramachandran, R.; Rodriguez, M. C. Student Perspectives on Remote Learning in a Large Organic Chemistry Lecture Course. *J. Chem. Educ.* **2020**, 97, 2565–2572.

10. Petillion, R. J.; McNeil, W. S. Student Experiences of Emergency Remote Teaching: Impacts of Instructor Practice on Student Learning, Engagement, and Well-Being. *J. Chem. Educ.* **2020**, 97, 2486–2493.

11. Jeffery, K. A.; Bauer, C. F. Students' Responses to Emergency Remote Online Teaching Reveal Critical Factors for All Teaching. *J. Chem. Educ.* **2020**, 97, 2472–2485.

12. Hurst, G. A. Online Group Work with a Large Cohort: Challenges and New Benefits. *J. Chem. Educ.* **2020**, 97, 2706–2710.

13. Gemmel, P. M.; Goetz, M. K.; James, N. M.; Jesse, K. A.; Ratliff, B. J. Collaborative Learning in Chemistry: Impact of COVID-19. *J. Chem. Educ.* **2020**, 97, 2899–2904.

14. Stowe, R. L.; Esselman, B. J.; Ralph, V. R.; Ellison, A. J.; Martell, J. D.; DeGlopper, K. S.; Schwarz, C. E. Impact of Maintaining Assessment Emphasis on Three-Dimensional Learning as Organic Chemistry Moved Online. *J. Chem. Educ.* **2020**, 97, 2408–2420.

15. DeKorver, B.; Chaney, A.; Herrington, D. Strategies for Teaching Chemistry Online: A Content Analysis of a Chemistry Instruction Online Learning Community during the Time of COVID-19. *J. Chem. Educ.* **2020**, 97, 2825–2833.

16. *Bachelor of Pharmaceutical Science, Monash Univeristy*. Available from: https://www.monash.edu/pharm/students/undergrad/course-information/bpharmsci (accessed 28 February, 2021).

17. White, P. J.; Larson, I.; Styles, K.; Yuriev, E.; Evans, D. R.; Rangachari, P. K.; Short, J. L.; Exintaris, B.; Malone, D. T.; Davie, B.; Eise, N.; Naidu, S.; McNamara, K. Adopting an active learning approach to teaching in a research-intensive higher education context transformed staff teaching attitudes and behaviours. *Higher Education Research & Development* **2016**, 35, 619–633.

18. McLaughlin, J. E.; White, P. J.; Khanova, J.; Yuriev, E. Flipped classroom implementation: A case report of two higher education institutions in the United States and Australia. *Computers in the Schools* **2016**, 33, 24–37.

19. White, P. J.; Larson, I.; Styles, K.; Yuriev, E.; Evans, D. R.; Short, J. L.; Rangachari, P. K.; Malone, D. T.; Davie, B.; Naidu, S.; Eise, N. Using active learning strategies to shift student attitudes and behaviours about learning and teaching in a research intensive educational context. *Pharmacy Education* **2015**, 15, 116–126.

20. White, P. J.; Naidu, S.; Yuriev, E.; Short, J. L.; McLaughlin, J. E.; Larson, I. Student Engagement with a Flipped Classroom Teaching Design Affects Pharmacology Examination Performance in a Manner Dependent on Question Type. *Am. J. Pharm. Educ.* **2017**, 81, 5931.

21. Lyons, K. M.; Brock, T. P.; Malone, D. T.; Freihat, L.; White, P. J. Predictors of Examination and Objective Structured Clinical Examination Performance in a Flipped Classroom Curriculum. *Am. J. Pharm. Educ.* **2021**, 85, 8038.
22. FutureLearn. *The Science of Medicines*. Available from: https://www.futurelearn.com/courses/the-science-of-medicines (accessed 28 February, 2021).
23. Manallack, D. T.; Yuriev, E. Ten Simple Rules for Developing a MOOC. *PLoS Comput. Biol.* **2016**, 12, e1005061.
24. *Panopto*. Available from: https://www.panopto.com/ (accessed 28 February, 2021).
25. *LibreTexts*. Available from: https://libretexts.org/ (accessed 28 February, 2021).
26. Trust, T. *The 3 Biggest Remote Teaching Concerns We Need to Solve Now*. Available from: https://www.edsurge.com/news/2020-04-02-the-3-biggest-remote-teaching-concerns-we-need-to-solve-now (accessed 28 February, 2021).
27. Hoon, A.; Oliver, E.; Szpakowska, K.; Newton, P. Use of the 'Stop, Start, Continue' Method is Associated with the Production of Constructive Qualitative Feedback by Students in Higher Education. *Assessment & Evaluation in Higher Education* **2015**, 40, 755–767.
28. Kaur, M. Blended learning - its challenges and future. *Procedia Social and Behavioral Sciences* **2013**, 93, 612–617.
29. Whitford, T. Let's not forget quality in our rush to deliver classes online, at scale. *Campus Review*. 1 May, 2020.

Chapter 9

Student Experiences and Perceptions of Emergency Remote Teaching

Barbara Chiu and Nicole Lapeyrouse

Department of Chemistry, University of Central Florida, 4000 Central Florida Boulevard, Orlando, Florida 32816, United States
*Email: nicole.lapeyrouse@ucf.edu

In this study, student responses to a survey about COVID-19 emergency remote teaching were analyzed to gauge student perceptions of the online transition. With the hasty shift to online, the student experience was different from typical online classes that were online for the whole semester. Several key positives and negatives emerged from coding the student responses. Currently, students report significant challenges in mental health, communication with peers and instructors, and balancing home and school life (especially finding a quiet environment to study). Loss of motivation, lack of face-to-face experiences, and faulty technology were also cited as major detriments to student learning, which can be taken into consideration when constructing courses in future semesters and can benefit students in the online environment. This illustrates how important physical experiences and resources are to teaching STEM courses. Especially in labs, being able to physically interact with the environment and talk to the professors and other students were crucial to student success. The information from this paper will help inform future decisions on online teaching and emergency remote teaching and can help educators adapt their classrooms for an enhanced online student learning experience.

Introduction

The spring semester of 2020 introduced unprecedented emergency remote teaching due to the worldwide COVID-19 pandemic. With the sudden shift to online, the student experience was very different from the typical semester. This paper will examine student perceptions of the rapid transition to emergency remote teaching, factors that affected these perceptions, and changes to student perception of online teaching after the remote teaching experience. Many instructors have reported changing their teaching methods due to the online transition and expressed a desire for more information on how to support student learning even while remote (1). These student perceptions may be important to improve teaching pedagogies for future emergency situations and improve institutional resilience as a whole (1).

It is important to note that emergency remote teaching is not the same as online teaching. Online teaching pedagogies have been researched extensively for decades and have flexibility around how much the class should be online, content delivery, instructor/student/peer interactions, and so on (2–4). These online classes were designed with the online interface in mind and carefully constructed to maximize student satisfaction and learning success (5–7). Emergency remote teaching was only made as a temporary solution to allow students to access instruction easily and reliably to finish the semester during an emergency situation (8).

Still, many valuable lessons can be learned from emergency remote teaching. Research in this field is still limited due to how recent spring remote teaching was, but the papers in the field have already noticed some common themes of student perceptions towards emergency remote teaching.

Literature Review

With the hasty shift to emergency remote teaching, the student experience was different from typical online classes. Overall, the unexpected transition and an unfamiliar, new way of teaching led to a largely negative student experience (9). Several key student concerns about emergency remote teaching have already emerged (9–11), although more research is still necessary to confirm these results with a wider pool and add to this discussion of student perceptions.

Academic Concerns

Among student academic concerns during the online transition, adjusting to more self-guided learning and different coursework/mode of teaching were among the most discussed topics. With remote learning, students have a greater responsibility to complete coursework and attend classes, as there is no longer any physical presence or event to push the students to complete their homework or participate in class. Many of the students surveyed said that motivation was one of the biggest challenges in emergency remote learning, especially balancing family, work, and school responsibilities (12, 13). This was especially true for lower-income and minority students as they were limited in their time to study and were forced to prioritize other obligations over classwork (14). A number of students depended on a study space away from home to be able to work and study for classes in a quiet environment without any distractions from family members or domestic issues (9, 12). Jeffery et al. describes these types of students as "structure-seeking," who focus on extrinsic structures (such as a set time and location for class meetings, reserving a study room at a certain time) to help frame where, when, and how to work (15). For these students, the unexpected shift to remote teaching led to decreased levels of concentration and productivity. As stated by one student, "My biggest challenge going into the new method of teaching was time management. Since I did not have to actually meet in a classroom, I had to make sure I was staying up to date with my work and turning my assignments in on time" (11). The transition to emergency remote classes was extremely difficult for students with low self-efficacy and who thrived in a learning environment with high engagement and interaction with the professors and students. In physical classrooms, the physical encouragement from classmates and even the presence of other people in the room drove students to participate in class (14). With classes now delivered through videos and other online formats, students began to view emergency remote classes as passively watching videos instead of a lively classroom environment (14). Participation in classes dropped significantly, due to decreased attention and difficulty utilizing technology. In particular, classwork delivered in emergency remote classrooms became extremely difficult to finish, and many features of the emergency remote learning platforms were cumbersome for the students (*14*, *16*). Other students had limitations with their technology, such as lack of Wi-Fi and having to share their devices with multiple household members (*10*, *14*, *16–18*). In one survey, approximately 30% of students reported that technology issues had significant impacts on their learning (*10*). A student stated, "I have 3 people trying to stream videos at once in my home so our wifi bandwidth cannot maintain everyone at once. I often get kicked off of some zoom meetings or find myself having to watch the recordings later when less people are using the internet" (*10*). These factors exacerbated the participation problems with emergency remote teaching and caused additional stress on students.

The changes in coursework and mode of teaching were also a great cause for concern among students. Several students reported difficulty working on group projects and research reports as they had less communication with classmates, librarians, or instructors (*12*, *13*). When the classes were in-person, it was extremely convenient to ask the instructor a question after class or stop by office hours. Many students did not take advantage of online channels of communication provided by instructors through email, phone, or online meetings (*14*). The drastic changes to the course structure and syllabus mid-way through the semester gave even more confusion to the students and added to their overall anxiety (*19*).

The loss of face-to-face experiences was also cited as a major challenge to students, especially in laboratory classes where hands-on experience is critical to learning (*12*). Lab work was extremely difficult to adapt to emergency remote platforms, and the students were disappointed to lose the practical aspects of the lab for data analysis assignments they would complete at home (*20*, *21*). Group work in the lab was nearly impossible to carry out effectively (*20*, *22*). Working remotely with other students was especially difficult when all the members needed to talk or complete an activity synchronously (*20*). While students expressed a need to actively participate and discus with their peers, they also desired an asynchronous mode of content delivery to complete their work on a flexible schedule (*9*, *12*, *17*, *22*). One student stated, "I am 3 h ahead and am often extremely tired or cannot attend office hours, etc. because of the time" (*10*). Many students stated that they believed instructors tried their best to ensure their success during the online transition; however, these factors ultimately led to mostly negative student perceptions of emergency remote teaching (*9*).

Non-academic Concerns

While many of the studies so far about emergency remote teaching focus on student academic concerns, it is also important to realize that many students are facing non-academic concerns during the pandemic that could negatively affect their learning. Approximately half of the students surveyed from universities across the U.S. reported feelings of anxiety, stress, and other mental and physical health concerns during the pandemic (*13*, *23*). In a survey, a student stated, "I have not left my home or interacted with anyone but my mother for the last 45 days or so. I am starting to get lonely and my depressive symptoms are beginning to return. I also cannot sleep at night" (*10*). The impacts of the COVID-19 pandemic affected every aspect of student lives and caused a great deal of stress on students, who were still expected to continue their studies and perform well academically. Many students also lost their jobs during the pandemic or had reduced hours/pay, which led to rapid increases in food or housing insecurity (*17*). Some surveys report one in every three students having concerns with basic needs insecurity during the pandemic, while others report numbers as high as 3 in 5 students and even higher for students who are caregivers for their children or family members (*13, 23, 24*). These stressors are exacerbated by the increased demand for self-reliance and independent learning in emergency remote teaching.

Research Purpose

In this study, student responses to a survey about COVID-19 emergency remote teaching were analyzed to gauge student perceptions during and after the online transition. There is a high demand by instructors and institutions for more information on how to improve teaching remotely, but the research on this topic is still restricted to a limited pool of papers (*1*). Research on improving teaching pedagogies for emergency remote teaching is important to better prepare instructors for future emergency situations and improve overall institutional resilience (*1*). Many of the challenges students identified in their emergency remote classrooms are pre-existing problems that have only been exacerbated with the online transition (*15*). The information from this paper will help inform future decisions on online remote teaching and emergency remote teaching and help educators adapt their classrooms for enhanced student learning even after this pandemic.

Method

Participants

The survey participants included 229 students enrolled in chemistry undergraduate and graduate courses at the University of Central Florida. Participants were recruited through e-mails that were distributed within the chemistry department and online announcements. All participants were volunteers and were not excluded from the study based on race, gender, or age (IRB ID: STUDY00001667).

Materials and Procedure

The survey was designed to better understand student experiences and measure the degree to which student perceptions of online courses were altered due to this transition. As the survey was sent online through Qualtrics, participants were free to take the survey at any time or location. The research purpose and explanation of how the results would be used was explicitly written at the beginning of the survey. The survey participants responded to several open response questions about aspects of online learning during spring 2020, such as accessibility or format content was delivered (see Appendix A for the complete survey questionnaire).

Methods

After the survey collection ended, the responses were read thoroughly, and several commonly repeated themes were identified. The codebook, as shown in Table 1, was created through emergent coding as codes were created from the responses after the data was collected, not prior to the analysis (*25*). The themes emerged from the student responses were grouped under large categories such as accessibility or health. The survey responses were then coded for these themes, as well as indications of whether the response was in a positive or negative tone. Positive tone responses discussed the benefits of emergency remote teaching, while negative tone responses discussed the drawbacks of emergency remote teaching. A second researcher coded the survey responses as well with a percent agreement of 0.90, indicating high levels of agreement for the codes. The frequency of the codes was then counted in Excel, and bar graphs were created from the data. The N value for each bar graph was reported as the number of students participating in the survey. This value differs from frequency of codes of that question because multiple codes could be assigned to each student response. In some cases, student responses did not provide enough information to code; for example, responses of "N/A" or "This wasn't a factor" were not included in the analysis.

Table 1. Codebook

Accessibility	Indicates discussion of learning accessibility (such as location and time flexibility) due to changes in the external learning environment.	"I used to commute 45min everyday so it's easier to work when I just have to go to my computer."
Communication	Indicates topics of communication, whether it be between students or with the professors. This may include announcements sent to the classes, responses through email, and other forms messaging and discussion.	"There were many changes made to syllabi but professors weren't responding in a timely manner."
Course Design	Refers to the organization of the course and how content was delivered. Also includes the transition from in-person to emergency remote classes.	"Some professors used the polling feature on zoom to encourage participation."
Health	Discusses the impacts of health during the transition to online. This includes how mental health may have made learning more difficult, or other health issues that played a role in student learning during the online transition.	"I had mental health issues that were exacerbated during this time."
Motivation	Deals with personal determination issues and work ethic. Also refers to how students have the motivation to develop routine schedules for studying and to adapt to the online transition.	"It's hard to find the motivation to make myself learn."
Resources	Refers to resources available to students to help them succeed in their classes, or resources students found online including videos, articles, practice problems, and more.	"My professors would post more worksheets."
Technology	Refers to anything dealing with the use of technology in the classrooms, such as various computer programs or meeting platforms.	"Internet is not always reliable especially when the entire country is using it."

Results and Discussion

Students Struggled with Course Organization and Engagement

Many of the student responses comparing transitioned classes to online classes had a heavy course design focus, mostly in a negative light (Figure 1). Their discussion was framed around physical components of the course, such as the organization of the material, how the course changed during the online transition, online incompatibility, and difficulty of the course material. For example, one student's response said, "My previous online courses were much better as the online format was already well established and organized, as compared to the courses that were hastily transitioned to online as a result of COVID-19." The hasty transition to online meant that the classes were not set up to be taught in an online format, so it was more difficult for students to find the course modules, assignments, and lectures. This was to be expected since face-to-face courses are not designed to be fully online. Students also believed that some classes, especially in STEM, required more instructor engagement and a learning environment that could only be offered in-person. Staying engaged and understanding the material became much more difficult, and there was a reported decrease in motivation with the transition to more independent learning. There was also greater importance on the availability of study materials and tutoring. Online tutoring, provided by the university, was cited as a major help to students who needed more personal study time to go over concepts and understand the material better. Online study materials were also greatly helpful to students, but it seemed that student experiences varied from professor to professor. While some students reported that their professors provided more online resources to help with the course material, others reported the opposite and cited a lack of resources as a cause for their learning to suffer. Overall, most of the students preferred classes that started online rather than those that transitioned.

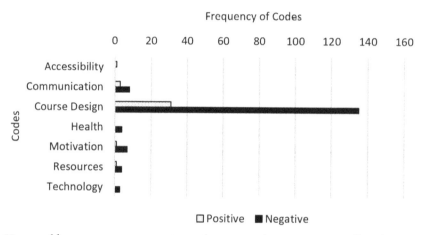

Figure 1. How would you compare your previous online courses that transition rapidly to being on online as a result of COVID-19? (n = 97)

Emergency Remote Teaching Created Limitations with Interactions

The transition to emergency remote teaching caused students to lose valuable in-person experiences, such as working in the lab, attending conferences, and even interacting with instructors and peers in class. The most common limitations students responded with were technological issues (including accessibility to technology, lack of technology), environment, loss of experience, and professor/TA communication. When discussing the transition of STEM courses, students placed a greater focus on course design and resources (Figure 2). This illustrates how important physical experiences and resources can be to teaching STEM courses. Especially in labs, being able to physically interact with the materials and talk to the professors and other students was crucial to the student experience. Students felt that they missed out on other face-to-face experiences as well, such as talking with professors or working in the lab and also remarked on the slower communication with professors. Several students reported that their class syllabus was modified due to the online transition, such as changes in exam format (time allotted, types of questions, etc.) and grading (less extra credit, dropping grades, etc.). However, these changes were not always communicated clearly with the class and caused confusion on grading and assignment details. Students who had questions about the material or course structure found it difficult to schedule meetings or receive responses from instructors, who were receiving large influxes of e-mails after the online transition. A lack of communication was one of the most significant drawbacks students reported in the online transition and contributed greatly to student feelings of disconnect from being in a physical classroom.

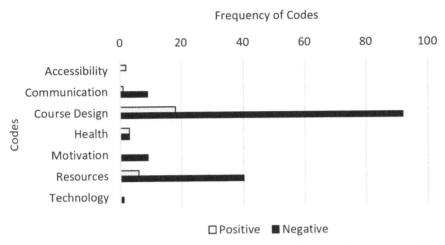

Figure 2. Traditionally, STEM courses are exclusively taught face-to-face. Has the transition of these courses to being offered online had any impact? (n = 97)

Emergency Remote Teaching Provided Students with Flexibility but Limited Access to Resources

There were also some key benefits of emergency remote teaching, namely the accessibility of these classes (Figure 3). Several student responses centered around the benefits of online transition, including schedule flexibility, study materials/tutoring, and less commute. An example of a positive student response is, "I've been able to plan my schedule with more flexibility since classes have moved to online. I can better manage my time now." Most students said that the transitioned emergency remote classes gave them more freedom to choose when and where to work/study due to the fact that all of the material was online and that there was no commute to classes anymore. Specifically, students commented on how they did not have to commute as much after transitioning online, saving them time and money. As one student stated, "It [college courses] being online ultimately removed 10 hours of commute time from my life, decreased an immense amount of stress from travel and gas prices." Students enjoyed working at their own pace and setting their own schedule with emergency remote classes, which allowed them to work ahead and then clear their schedule for personal free time or to help their families. Many students also commented on the fact that emergency remote classes have greater student capacity, so they were able to take more classes and easily enroll for the summer semester. However, the heavy dependence on technology also caused many issues with accessibility (Figure 4). Specifically, student responses centered around resources and technology, such as loss of experience, study materials/tutoring, and technological issues (Figure 4). One negative student response stated, "I also had a lot of internet access problems and it would go out during quizzes, lecture, and studying." Furthermore, several students felt that there were not as many resources provided to them when they moved online and that they lost out on valuable in-person experiences (such as working in the lab and interacting with peers). The technology code was also commonly mentioned, as students had issues acquiring expensive technology (such as webcams) and using the programs in an online classroom. Many students struggled with poor Wi-Fi or having no computers at home, so they were unable to access everything provided online. As mentioned before, the pandemic has already cut the wages of many students, so having to purchase computers or webcams to take tests and work on coursework was very stressful for some students. Many institutions do have programs to loan laptops and webcams to students, but they were closed during the pandemic and students were unable to utilize these campus resources. The student responses regarding accessibility highlights the issues that disadvantaged students faced during the online transition and how this could have impacts on their learning.

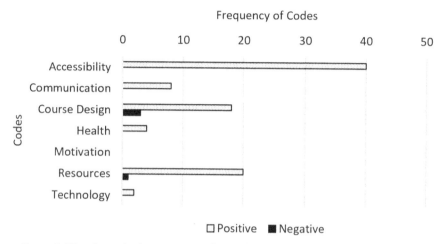

Figure 3. How has it (online transition classes) been made more accessible? (n = 97)

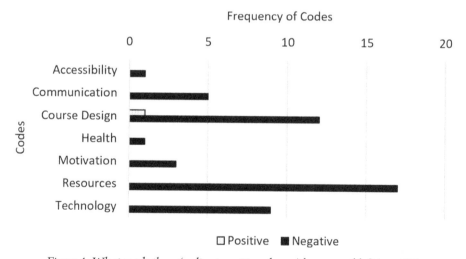

Figure 4. What made them (online transition classes) less accessible? (n = 97)

External Factors Impacted Students' Abilities to Complete Assignments

The discussion of online student experiences elicited many more responses about students' personal lives and external factors affecting their education (Figure 5). There were still discussions of course design factors, such as pacing, grading scale, and exam format, but this was minor compared to the external factors mentioned below. Course design remained the most common primary code, but there was an increase in health and motivation primary codes as well. Specifically, students discussed the impacts of the online transition on the course structure, such as having to learn more material independently. Unlike in-person classes, with set times and locations to keep students accountable, online lectures had less pressure on students to attend and participate. Some instructors began relying more heavily on textbooks and assigning reading material or videos to deliver instruction to students, which requires more motivation and independent learning from the students to succeed. More responses also considered the health primary code during the online transition, whether it was mental health or other health issues that affected their ability to focus on their class. For example, one student stated, "I had mental health issues that were exacerbated during these times and it was hard to focus on my schooling by not being immersed in the campus." Students struggled to adjust to a much different learning environment and relearn many of their previous study methods. This adjustment was particularly difficult for students dealing with personal issues, such as a loss of money or family conflicts, that made it hard for them to fully focus on classes at the same time. Several surveys have found that many students are struggling with basic need insecurities, such as food or housing, due to wage-cutting or losing jobs during the pandemic (23, 26). These factors outside of the classroom play an equally significant role in student learning, and are partially responsible for the stress and decreased motivation of students during this time.

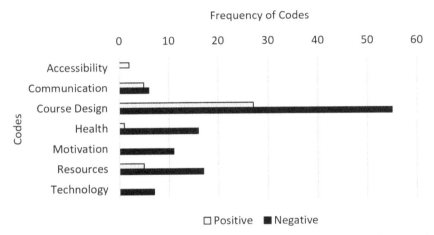

Figure 5. How would you describe your online experience during spring 2020? (n = 97)

Conclusion

The student responses focused on different aspects of course design and often had a negative outlook. Many students believed the emergency online transition caused the course formats to be disorganized and unstructured, and this made it difficult to find assignments and learning materials for each chapter or module of the course. Another issue students discussed was the course material itself. Especially for STEM classes such as organic chemistry or physics, students commented that the course material was not effectively designed for online formats and many students felt lost and confused with how to adapt to the online transition. This was further exacerbated by the lack of communication between instructors and students once emergency remote teaching was implemented. For future instances of emergency remote teaching, instructors may wish to focus on making the course simple to access and explain any changes to the course. Direct and quick communication between instructors, students, and their peers would also help students feel more engaged in the classroom and improve their learning.

Furthermore, accessibility and technology were two primary codes that were significant in many student responses. Students mentioned that a major difficulty with the emergency remote classrooms was the technological issues using zoom or other forms of content delivery. The costs of technology were also a detriment to student learning. However, there were some benefits to using technology, namely being that they increased the accessibility of courses. The emergency remote classrooms had much greater time and location flexibility, which students enjoyed. For students with jobs or a busy schedule outside of their education, the online transition was beneficial. This may be a positive sign for institutions to expand the range of course modalities offered in the future so that students with many extracurricular activities are able to work on their courses at their own schedule.

Motivation was another primary code from the responses. With classrooms transitioning to remote teaching, many of the responsibilities of learning fell onto the students. Therefore, many students with low self-motivation had little incentive to go to classes or actively participate while they were in these emergency remote classes. Students reported having to seek out additional resources or tutoring in their free time so they could understand the course material. It is important for instructors to be accommodating to students in future emergency remote classes and possibly institute more participation checkpoints to hold students accountable and actively engaged in the classrooms (such as questions to answer during the lectures or counting attendance).

The transition to emergency remote classes in the spring was an unprecedented move due to the COVID-19 pandemic, and with the transition came many difficulties for students and instructors alike. Future transition classes should try to implement greater communication between instructors and students, more structure in the online classroom settings, and less emphasis on independent learning, while institutions may consider putting programs in place to help financially struggling students and expanding the availability of online classrooms for students with busier lives once courses return back to normal.

Appendix A. Survey Questionnaire

Q1 - When courses were offered face-to-face did you commute to campus? (Y/N)

- (If Yes is selected) When course were offered face-to-face did you commute to campus?
- (If Yes is selected) About how long was the commute?
  Less than 10 minutes
  10-20 minutes
  20-30 minutes
  30-60 minutes
  Other (please explain)

Q2 - Have you ever taken an online class before? (Y/N)

- (If Yes is selected) What online course have you taken before?
- (If Yes is selected) How would you compare your previous online courses to courses that transition rapidly to being on online as a result of COVID-19?

Q3 - How would you describe your online experience during spring 2020?
Q4 - What courses were you taking during spring 2020?
Q5 - Traditionally, STEM courses are exclusively taught face-to-face. Has the transition of these courses to being offered online had any impact?
Q6 - Has the availability of courses being offered online made them more accessible?

- (If Yes is selected) How has it been made more accessible?
- (If No is selected) What made them less accessible?

Q7 - What aspect of taking these courses did you find beneficial or detrimental?

Q8 - How did your course change as a result to shifting online?

Q9 - How was your content being delivered during spring 2020? (Please elaborate on any lecturer videos that were provided, audio recording, and/or online activities)

Q10- Were there any limitations that you experienced during spring 2020?

Q11- How was your performance in the course and understanding of the content impacted due to the change?

Q12- What were the benefits for your courses being move to online?

Q13- What problems or difficulties did you experience using new technology?

Q14- What is the likelihood of you taking an online course again once face to face classes are offered?

Very likely

Moderately likely

Neutral

Moderately unlikely

Very unlikely

References

1. Johnson, N.; Veletsianos, G.; Seaman, J. U.S. Faculty and Administrators' Experiences and Approaches in the Early Weeks of the COVID-19 Pandemic. *Online Learn.* **2020**, *24*, 6–21.
2. Hodges, C. B.; Moore, S.; Lockee, B. B.; Trust, T.; Bond, A. The Difference Between Emergency Remote Teaching and Online Learning. *Educause Rev.* **2020**.
3. Wallace, R. M. Online Learning in Higher Education: a Review of Research on Interactions Among Teachers and Students. *Educ. Commun. Inf.* **2003**, *3*, 242–280.
4. Keengwe, J.; Kidd, T. T. Towards Best Practices in Online Learning and Teaching in Higher Education. *J. Online Learn. Teach.* **2010**, *6*, 533–541.
5. Nortvig, A.; Petersen, A. K.; Balle, S. H. A Literature Review of the Factors Influencing E-Learning and Blended Learning in Relation to Learning Outcome, Student Satisfaction, and Engagement. *Electron. J. e-Learn.* **2018**, *16*, 46–55.
6. Castro, M. D. B.; Tumibay, G. M. A Literature Review: Efficacy of Online Learning Courses for Higher Education Institution Using Meta-analysis. *Educ. Inf. Technol.* **2019**, *26*, 1367–1385.
7. Fisher, M.; Baird, D. E. Online Learning Design that Fosters Student Support, Self-regulation, and Retention. *Campus-Wide Inf. Syst.* **2005**, *22*, 88–107.
8. Bozkurt, A.; Sharma, R. C. Emergency Remote Teaching in a Time of Global Criss due to Coronavirus Pandemic. *Asian J. Distance Educ.* **2020**, *15*, 1–6.
9. Petillion, R. J.; McNeil, W. S. Student Experiences of Emergency Remote Teaching: Impacts of Instructor Practice on Student Learning, Engagement, and Well-Being. *J. Chem. Educ.* **2020**, *97*, 2486–2493.
10. Ramachandran, R.; Rodriguez, M. C. Student Perspectives on Remote Learning in a Large Organic Chemistry Lecture Course. *J. Chem. Educ.* **2020**, *97*, 2565–2572.
11. Simon, L. E.; Genova, L. E.; Kloepper, M. L. O.; Kloepper, K. D. Learning Postdisruption: Lessons from Students in a Fully Online Nonmajors Laboratory Course. *J. Chem. Educ.* **2020**, *97*, 2430–2438.

12. Vielma, K.; Brey, E. M. Using Evaluative Data to Assess Virtual Learning Experiences for Students During COVID-19. *Biomed. Eng. Educ.* **2020**, *1*, 139–144.
13. Blankstein, M.; Frederick, J. K.; Wolff-Eisenberg, C. Student Experiences During the Pandemic Pivot. *Ithaka S+R.* **2020**; DOI: 10.18665/sr.313461.
14. Kalman, R.; Macias Esparza, M.; Weston, C. Student Views of the Online Learning Process during the COVID-19 Pandemic: A Comparison of Upper-Level and Entry-Level Undergraduate Perspectives. *J. Chem. Educ.* **2020**, *97*, 3353–3357.
15. Jeffery, K. A.; Bauer, C. F. Students' Responses to Emergency Remote Online Teaching Reveal Critical Factors for All Teaching. *J. Chem. Educ.* **2020**, *97*, 2472–2485.
16. Soares, R.; de Mello, M. C. S.; da Silva, C. M.; Machado, W.; Arbilla, G. Online Chemistry Education Challenges for Rio de Janeiro Students during the COVID-19 Pandemic. *J. Chem. Educ.* **2020**, *97*, 3396–3399.
17. Kimble-Hill, A. C.; Rivera-Figueroa, A.; Chan, B. C.; Lawal, W. A.; Gonzalez, S.; Adams, M. R.; Heard, G. L.; Gazley, J. L.; Fiore-Walker, B. Insights Gained into Marginalized Students Access Challenges During the COVID-19 Academic Response. *J. Chem. Educ.* **2020**, *97*, 3391–3395.
18. Kyne, S. H.; Thompson, C. D. The COVID Cohort: Student Transition to University in the Face of a Global Pandemic. *J. Chem. Educ.* **2020**, *97*, 3381–3385.
19. Murphy, L.; Eduljee, N. B.; Croteau, K. College Student Transition to Synchronous Virtual Classes during the COVID-19 Pandemic in Northeastern United States. *Pedagog. Res.* **2020**, *5*.
20. Dietrich, N.; Kentheswaran, K.; Ahmadi, A.; Teychené, J.; Bessière, Y.; Alfenore, S.; Laborie, S.; Bastoul, D.; Loubière, K.; Guigui, C.; Sperandio, M.; Barna, L.; Paul, E.; Cabassud, C.; Liné, A.; Hébrard, G. Attempts, Successes, and Failures of Distance Learning in the Time of COVID-19. *J Chem. Educ.* **2020**, *97*, 2448–2457.
21. Dickson-Karn, N. M. Student Feedback on Distance Learning in the Quantitative Chemical Analysis Laboratory. *J. Chem. Educ.* **2020**, *97*, 2955–2959.
22. Rodríguez-Rodríguez, E.; Sánchez-Paniagua, M.; Sanz-Landaluze, J.; Moreno-Guzmán, M. Analytical Chemistry Teaching Adaptation in the COVID-19 Period: Experiences and Students' Opinion. *J. Chem. Educ.* **2020**, *97* (9), 2556–2564.
23. Goldrick-Rab, S.; Coca, V.; Kienzl, G.; Welton, C. R.; Dahl, S.; Magnelia, S. *#RealCollege During the Pandemic.* https://hope4college.com/realcollege-during-the-pandemic/ (accessed 7/2/20).
24. Soria, K. M.; McAndrew, M.; Horgos, B.; Chirikov, I.; Jones-White, D. Undergraduate Student Caregivers' Experiences during the COVID-19 Pandemic: Financial Hardships, Food and Housing Insecurity, Mental Health, and Academic Obstacles. *SERU Consortium Reports* **2020**.
25. Stemler, S. An Overview of Content Analysis. *Pract. Assess. Res. Eval.* **2000**, *7*, 17.
26. Blankstein, M.; Wolff-Eisenberg, C.; Braddlee, B. Student Needs Are Academic Needs: Community College Libraries and Academic Support for Student Success. *Ithaka S+R.* **2019**; DOI: 10.18665/sr.311913.

Chapter 10

Students as Partners: Co-creation of Online Learning to Deliver High Quality, Personalized Content

Amy L. Curtin[1] and Julia P. Sarju[2,*]

[1]Independent scholar, graduate of the Department of Chemistry, University of York, Heslington YO10 5DD, United Kingdom
[2]Department of Chemistry, University of York, Heslington YO10 5DD, United Kingdom
*Email: julia.sarju@york.ac.uk

Students have valuable insights into their own learning experiences and preferences, and as such are well placed to inform curricula and learning design, ensuring that their needs are met. Collaborative pedagogies including co-creation and students-as-consultants can increase student engagement leading to improved student feedback, attainment, and retention. Students' learning experiences and needs are diverse; active steps must be taken to ensure that all students' voices are heard and that partnerships are inclusive of traditionally minoritized students. Working in partnership with students can further equality, diversity, inclusivity, and respect (EDIR) goals. This can be achieved when used to raise the voices of and empower minoritized students; developing culturally relevant and inclusive curricula which meet real rather than perceived needs. Although there has been a wealth of studies of collaborative pedagogy projects with students in higher education, there are relatively few reports of this approach in chemistry education. This chapter explores student-instructor pedagogical partnerships in online learning in chemistry education and is informed by wider educational literature, including evaluations of partnerships in other disciplines. We argue that the co-creation of programs, modules, learning materials, and assessments with student partners is an effective framework for learning design, which has the potential to improve inclusivity, engagement, and attainment. Benefits have been reported for both instructors and students contributing to pedagogical partnerships. It appears that this approach is underused in chemistry education and we believe that further development in this area would lead to greater insights into its effectiveness within chemistry education, further EDIR goals, and empower student voices. However, without careful consideration of how student partnerships are executed, inequities could be further entrenched.

© 2021 American Chemical Society

Introduction

Traditionally, higher education instructors designed curricula and materials independently from their students, usually collecting feedback after the delivery of a course to inform future delivery (1). Student-instructor partnerships access the insights of students into their own needs, and diverse perspectives of learners, in addition to the instructors' perceptions of needs, throughout the design and delivery of courses (2). The incorporation of student perspectives into the design of chemistry education could be highly beneficial, where high-level concepts that require several levels of understanding simultaneously are commonplace (3, 4), increasing cognitive load (5, 6).

A small, but growing area of literature combines online chemistry education with co-creation: the development of curricula and course materials through collaboration between instructors and students, centering learner empowerment and influence. Through co-creation, both instructors and students have a voice and a stake in the outcomes of the collaboration (1, 7). Beyond chemistry, co-creation pedagogies have been widely adopted (1), and there are dedicated journals for reporting scholarship of teaching and learning developed in partnership with students (8, 9). The level of student participation in education design can be modeled as a ladder (see Figure 1), moving from a 'dictated curriculum' developed without student involvement to 'student led' where students are in control of their curriculum (10). As students' level of participation increases, their role in education design transforms from User, to Tester, then Informant/Consultant, to Design Partner. The level of student participation refers to the level of student engagement, choice, influence, and the areas of the curriculum within which these apply.

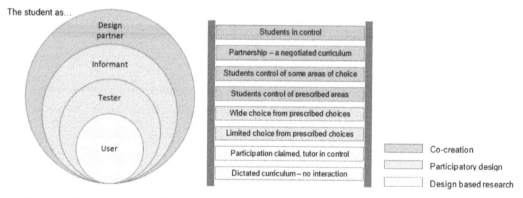

Figure 1. The onion model (left) of student roles in education design and ladder model (10) (right) of student participation. Reproduced with permission from reference (7). Copyright 2019 Taylor & Francis Ltd. Access online at www.tandfonline.com. [Transcription: onion model of student roles, widening from user, tester, informant to design partner. Ladder model of student participation increasing from: dictated curriculum – no interaction, participation claimed – tutor in control, limited choice from prescribed choices, wide choice from prescribed areas, student control of prescribed areas, student control of some areas of choice, partnership – a negotiated curriculum, to students in control.]

The adoption of online learning and teaching pedagogies in chemistry higher education, which has been increasing over the last decade (11, 12), has been dramatically accelerated by the Covid-19 pandemic (13, 14). Implementation of online learning is diverse and includes massive open online courses (MOOCs) (15), pre-laboratory exercises and simulations (16), online laboratory learning (17–20), augmented or virtual reality (21–26), synchronous and asynchronous lectures (27–30), flipped learning (31, 32), digital textbooks (33–35), discussion forums (28, 36–39), and digital tests

(*40*) such as multiple-choice quizzes (*41*). In the past, chemistry students have reported low levels of satisfaction with and enthusiasm for online learning. Their key concerns related to student experience (specifically enjoyability and perceived effectiveness of online teaching), instructional quality, and the ability of students to connect with faculty (*42, 43*). Designing and delivering materials for online teaching may also be unfamiliar to instructors. Instructors may lack the digital and accessibility skills required to produce online content suitable for all students (*44*). Co-creation can help to overcome these challenges, tapping into students' digital skills and preferences for (or familiarity with) certain platforms and teaching methods.

Strengths of online chemistry education include increased flexibility of asynchronous delivery, opportunities for the development of transferable digital skills, and the implementation of digital communication tools to facilitate student-student and student-instructor dialogue and contribute to fostering inclusive learning communities. Instructors can also use a host of chemical modeling software to enhance the teaching of challenging concepts such as 3D visualizations of chemicals, reactions, and processes (*44, 45*). Open educational resources such as MOOCs and open access digital textbooks can supplement formal learning and have the potential to democratize education, widen participation and encourage lifelong learning (*35, 46, 47*). In addition to addressing the concerns reported by students and instructors, online content co-produced with student partners can optimise the benefits.

This chapter reflects upon examples of co-creation in higher education, focusing on undergraduate students, both in online chemistry education and more widely. The focus is on pedagogical partnerships that collaborate on the theory, content, and overall design of curricula. Student representatives (*48*), student-led teaching delivery (*49*), students as partners in outreach work (*50*), and students as researchers where undergraduates undertake standalone or course-based undergraduate research experiences (CUREs) (*11*) will not be discussed in this chapter.

The aims of this chapter are to:

- Review the design and implementation of higher education online learning and teaching in chemistry by mixed collaborative teams of students and academic staff in higher education institutions;
- Explore different approaches to co-creation, different levels of student participation, and highlight examples of good practice;
- Consider how co-creation of curricula and/or learning materials affects student and instructor engagement or experience of teaching and learning;
- Explore whether adopting co-creation pedagogies can further Equality, Diversity, Inclusion, and Respect;
- Reflect upon ways in which mixed teams collaborate and consider the balance of power and equity of opportunities;
- Recommend strategies for effective, inclusive, and equitable collaboration in mixed teams including discussion of ethics, power distribution, credit, responsibility, and dissemination.

In this chapter, the term 'students' refers to students taught in formal higher education, primarily undergraduate chemists, and the term 'instructors' refers to any person traditionally involved in the design and delivery of this teaching. The term 'instructor' is inclusive of students who are employed to teach, often postgraduate students/doctoral candidates who may deliver and design curricula without being 'faculty', hence the distinction.

Author Positionality

In keeping with the students as partners approach this chapter has been written by authors with experience of taking part in collaborative pedagogic partnerships: one as a student and one as an instructor. Amy Curtin is a former student of chemistry at the University of York, now a Regulatory Scientist at the Health and Safety Executive. Co-creation formed an exciting part of her undergraduate education, through which she supported and encouraged the adoption of accessible online teaching practices and advised chemistry faculty. Julia Sarju is an assistant professor in Chemistry who has collaborated with students on a number of scholarship of learning and teaching projects, and in addition to being highly rewarding, these partnerships have been beneficial to her personal and professional development.

Case Study: Amy Curtin

As a chemistry undergraduate student at the University of York, I was recruited through a competitive internal employment selection process, as an 'e-Accessibility' intern within my own teaching department. I worked alongside faculty, learning program design specialists, and other students, across departments, over two consecutive summers (outside of semester). I came into this project with experience of disability, both lived and from close friends and family, but little awareness of the process of curriculum creation.

I was recruited as part of an institution-wide project to update online learning to meet new legal guidelines of proactive accessibility. All current and future online materials needed to display their core information so that a student with any disability could access it easily. My lead faculty contact was aware of the need for specialist help in the Department of Chemistry due to several factors: the large number of resources; the proportion of content displayed visually in figures and graphs; and the spread of content over several platforms and programs.

I worked part-time (30 hours per week) and was paid a living wage. Even before the Covid-19 pandemic, I was able to work remotely. This was an immense help as I was able to flexibly work from my out-of-semester address where I had other family commitments. I was trained in creating accessible materials when I began my role and collaborated at organized lunches and in online forums with student partners from other departments to share knowledge and solve problems together. However, I was given no formal pedagogical training and as my work progressed, I researched many emerging questions myself or discussed them with team members.

My work was split between consulting on the accessibility of new or existing resources with faculty members and researching and creating guides specific to teaching online chemistry accessibly. I discussed accessibility and universal design in teaching with faculty members who were highly engaged. My experience of co-creation falls outside traditional categories; I was a student co-creating curriculum materials with staff, but I was also advising staff as a non-subject matter expert. It was a high participation project, and I was the only student in this co-creation project from my department. This was managed well, however, as I rarely met with more than one or two faculty members at a time, so I did not feel intimidated, or as if I was at a job interview! Interactions with faculty members, especially face-to-face, made it clear I was a colleague, if a junior one, for example making small talk and eating lunch together.

There were proactive instructors who came to me for advice; this made me feel respected not just in my work but more generally as a part of the department. It raised my opinion of the department as an inclusive community. This sense of community applied across the university too, through my collaborations with other like-minded students. Through the discussions with these instructors, I influenced the pedagogy of creating accessible resources with students' inputs from the start, which

I hope improved future approaches to the creation of new modules. In a way, I co-created resources on co-creation, by encouraging instructors to involve disabled students in their course creation. Accessibility and co-creation are intertwined in this way.

The experience was definitely positive for me, as a student. I greatly improved my digital skills, learning a lot of short-cuts as I went along. Equally, my understanding of my chemistry course improved. As I reviewed and edited a wider array of resources than I had been taught with originally, I gained a new understanding of the core concepts. I had to improve my comprehension of all aspects of the course, to make sure when I created or changed materials, I wasn't altering the meaning.

In one of my guidance documents to faculty, I wrote *"Remember, the student knows best; it is their disability and their studies you are modifying"*. I think this applies to both accessibility and co-creation. Some of the edits I chose to make to course materials were only possible because of my expertise as a student, and as a student with lived experience of disability. Other students will have their own expertise to offer when you listen.

Collaborative Design and Creation of Online Learning and Teaching in Chemistry

Motivations to Adopt Co-creation Pedagogies

The motivations of co-creation online are broad. The overall aim, as in traditional curriculum creation, is to achieve high quality teaching, to improve learning (2). It is beneficial for higher education departments to be informed by students' insights into their learning, perceptions of effective strategies, and prior knowledge at the point of creation rather than after a course has been delivered (1). For instance students can provide guidance on effective strategies for learning with disabilities, which topics are interesting and culturally relevant to them, and their experience of learning technologies and platforms (digital skills).

Another common motivation of co-creation is increasing student motivation and engagement (51, 52). Many students report increased motivation and responsibility for their own learning after co-creating curricula (53), something which can be challenging for students moving from high school classroom education to more self-directed higher education online (54). Many projects also report being motivated by improving student-instructor relationships and communication (53). Minoritized STEM (Science, Technology, Engineering, and Maths) students report lower levels of belonging compared to their peers (55, 56), co-creation could be harnessed as a method to increase students' sense of belonging, and in the process improve motivation and attainment (57).

Levels of Participation

Levels of student participation and the amount that student input directly impacts the curriculum varies significantly across reports (1). Examples of co-creation in the literature can be broadly split into two categories: high participation projects with a small group of students, and low participation projects involving large proportions of a cohort (common examples are illustrated in Figure 2).

Co-creation projects may include but are not limited to:

- Co-compiling reading lists;
- Co-evaluating courses;
- Students and instructors choosing assessment methods;
- Co-grading assessments;

- Students and instructors designing curricula;
- Or co-creating course materials.

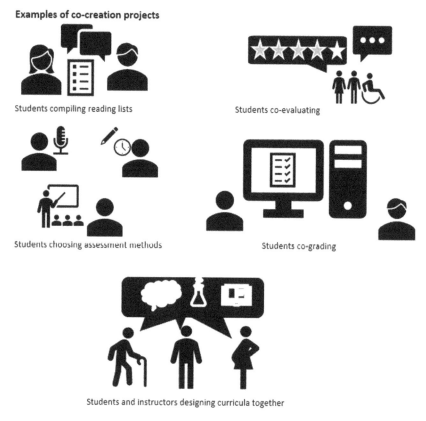

Figure 2. Pictorial representations of examples of student participation in curriculum and learning design. Five examples are given at the level of student empowerment we define as co-creation: students compiling course reading lists with instructors, students co-evaluating curriculum with instructors, students choosing their own assessment methods with no restrictions, students co-grading assessments, and finally students collaborating with instructors to design curricula. Two examples of student participation are given: students giving feedback and student representatives feeding back to instructors.

An example of a high participation project in online chemistry education is students as design partners in a negotiated curriculum. This was the approach taken in Goff and Knorr's co-development of a science curriculum through an Applied Curriculum Design framework (see section Building Curricula Together) (58). An example of a low level of participation is students as testers and informants within prescribed areas. This level of participation was reported by Aricò and Lancaster where the co-creation of multiple-choice questions using Peerwise was introduced in to an undergraduate inorganic chemistry course (see section Whole-cohort approaches) (41).

The influence of students on a pedagogical level often differs. Even examples involving high student participation may result in students having minimal input into the curriculum and learning objectives, for example, projects where students create materials as directed by instructors. These types of projects are not considered co-creation. In not working in partnership with students, opportunities for engagement and learning are likely to be lost (59).

Pedagogical Partnerships: What Are the Challenges?

Common challenges identified by Mercer-Mapstone *et al.* in a systematic review of reported examples of co-creation involving 11 case studies from STEM departments were (53):

- resistance to change by students, instructors, and institutions;
- institutional resistance to student involvement/partnership;
- equity of opportunity to participate (*e.g.* time restraints on instructors and students);
- and lack of experience of students in creating and designing courses.

Könings *et al.* identified further challenges (2):

- existing power imbalances between students and instructors;
- voice fatigue of students;
- and feelings of insecurity in instructors related to "*giving up control*" of courses.

Mercer-Mapstone *et al.* noted that in almost all of the successful examples of co-creation in the literature, the higher education institution was supportive of the individual co-creation project and the concept of students as partners to instructors rather than passive consumers (53). To support and value student participation, resources and funding are required by instructors and hence a supportive institution is beneficial to the success of co-creation projects (2). Some instructors mentioned requiring the final say on curricula, this is recommended for several reasons. Namely, that instructors, due to lower turnover rates than students, will retain the ownership of a curriculum and overall responsibility for its quality (60).

There are several barriers and perceived barriers to engaging students as partners and sharing pedagogical control. Instructors naturally wish to retain ownership of teaching and co-creation can feel threatening to this ownership, and harm instructors' morale, and/or make them resistant to co-creation projects (60). Additionally, high levels of student control over curricula can threaten instructors' (often already restricted) ownership in creating their curriculum, which does not lend itself to productive teamwork (53, 60).

Higher education institutions have a responsibility to students to provide high-quality education. This responsibility remains even where there are high levels of student participation. Students rarely have experience in designing and creating curricula prior to starting a project, and so may require support that instructors would not. It can be difficult for students to move from the lower participation mindset of simply evaluating courses to a higher level of participation designing solutions to problems and gaps in curricula (2). Co-creation should involve giving power and influence to student voice, but the responsibility for the curriculum quality remains with the instructor(s) (61).

Another key challenge is ensuring the whole student voice is actively listened to, not just a select few (*62–64*). We must be mindful that co-creative partnerships may prioritize the voices of students who are already privileged and most able to take up these opportunities (*65–67*). Mercer-Mapstone *et al.* warn that unless we actively ensure partnerships are inclusive, already marginalized students and instructors will be left out from the reported benefits (*53*).

Examples from Practice: Co-creation of Teaching and Learning in Higher Education

The types of curricula and the materials designed through co-creation are as varied as the ways of working involved. Martens *et al.* drew together five criteria for successful co-creation, based on examples from the literature, as defined in Table 1 below (*68*). These criteria are useful as a tool for evaluation and reflection of co-creation projects. In this section we consider student-created course materials, co-created curricula, and approaches involving whole cohorts as opposed to a sample of the student population.

Table 1. Definitions of the five criteria for successful student-instructor partnerships[a]

Criterion	*Definition*
Reciprocal respect	Taking each other seriously, valuing each other, and exchanging thoughts in an equal manner
Influence	Feeling that you can actually contribute to educational improvement
Autonomy	Determining by yourself how you will do your job
Commitment	Being willing to put extra effort into improving education and being worried about its quality
Ownership/ responsibility	Feeling that it is your job to improve the educational system

[a] Reproduced with permission from reference (*68*). Copyright 2019 Taylor & Francis Ltd.

Student-Created Course Materials

Chemistry, in all its forms, is a heavily visuospatial subject and chemistry students must learn, understand, and manipulate many 2D visual representations of this 3D world in any given task (*45*). Many misconceptions and difficulties can arise when using digital 2D representations of these concepts. Online resources created by other students, who have only recently overcome these difficulties, can improve students' understanding of theory. Furthermore, these digital representations can be difficult to convert to understandable alternative formats for students with visual impairments, input from students with relevant lived experience of disability, can be invaluable in the design of inclusive teaching resources (*69, 70*).

Recent successes include several projects where student-created videos formed part of the course materials (*52, 71, 72*). In these examples, student partners' perceptions of misconceptions and key points were shared with the next 'generation' of students, via online videos. For example, Gupta and Nikles selected student volunteers to create videos addressing a common student 'misconception' related to organic reaction mechanisms (*52*). The videos were then shared with the whole cohort prior to a laboratory experiment based on the same reaction. Students were surveyed after viewing the videos and completing the laboratory exercise; they found the assignment "*useful, enjoyable, and important*" (*N*=165). The students were given the freedom to design the videos as they saw fit, the authors reported that this led to the development of resources that addressed a wider array of learning preferences and methods than a single, instructor-created resource could.

Hubbard *et al.* report a recent success with student-created digital tutorials developed to support peer learning during an introductory undergraduate organic chemistry course (*51*). This co-creation project was tied directly to the course and was available to the whole cohort to take part. Student uptake of this voluntary activity was high (94%), and a wide variety of materials were produced including videos animated with the videogame Minecraft, cross-stitched visual aids, websites, and infographics. Students did not receive remuneration for this project, as the aim was to support student participant learning rather than to create resources. These resources are currently available publicly, and now form part of the course as review materials. Students taking part ($N=171$) reported increased understanding of organic chemistry concepts on the course (85%) and improvement of technical (31%) and study skills (35%). This example demonstrates the potential of co-creation for learning, not just within the co-created course but beyond to other courses, by supporting the development of transferable skills.

Many educators have harnessed the benefits of technology-enhanced teaching, working with student partners is a way to achieve this, learning from the experience of students, and also, as demonstrated above, their creativity (*44, 62*). The students in these examples also developed transferable skills and valued graduate attributes such as creative problem solving, independent learning, communication, flexibility, and digital skills that are considered lacking in chemistry graduates (*44*).

Whole-Cohort Approaches

Hwang adapted the digital card tool Plickers for remote, online, formative assessment in an undergraduate non-major chemistry course, in response to the Covid-19 pandemic (*73*). The platform was used to adapt the course delivery by altering the emphasis on different concepts depending on how well students performed on the associated Plickers questions. An average of 112 responses (from a cohort of 117 students and a total of 1,344 responses) were received for each online assignment, mainly coming from the same third of the students. Areas of poor performance were subsequently addressed in synchronous question and answer sessions where students could vote on the questions they wanted to be covered by instructors. This is an example of low-level student participation as 'testers' of the curriculum, able to influence the course content from a series of prescribed choices. It enabled the instructors to create resources appropriate for their students' needs, rather than creating resources based on preconceived notions that may have been less suited to student needs.

A further example of a whole-cohort approach was implemented by Aricò and Lancaster involving the co-creation of multiple-choice questions used for peer instruction to support an inorganic chemistry module (*41*). They provided their students with the stem of a multiple-choice question and then asked them to provide possible responses, using the online platform, Peerwise. This way, students provide their own misconceptions, in addition to those instructors know from experience of teaching the content. These co-created questions, answers, and comments on questions were entered into an online repository where they are available for students to discuss and review. Student feedback on this project was mixed, with positive comments reporting perception that the style improved engagement and improved learning outcomes, but also some negative comments where students reported finding the unfamiliar pedagogy difficult to adapt to. The positive feedback regarding the effectiveness of the pedagogical approach were supported by the measured improvement in attainment reported by Aricò and Lancaster.

Co-creation of Curricula

Goff and Knorr engaged students as genuine curriculum design partners, in a holistic way, for a foundational science course offered to first-year students (74). Students were recruited to a Curriculum Design in Science committee, which subsequently surveyed 324 upper-level science students as to their perspectives and needs of their foundational training. A subgroup of instructors and students then designed the core elements of the course, and subsequent small mixed teams created the resources forming these modules, for credit as part of their degree program. The project had the positive outcomes of increased and continuing student engagement; the project has been repeated every two years since its initial implementation with high uptake from students and instructors. Additionally, they report increased institutional support for co-creation, based on the success of the initial project.

Woolmer *et al.* also reported a case study where students collaborated with instructors to build an interdisciplinary foundational science course; this is an example of genuine co-creation of a curriculum (75). The whole cohort was surveyed to determine the topics students wished to be covered by the course, then students were invited to design the curriculum from the ground up. Instructors explicitly set traditionally off limit areas such as assessments firmly within the students' influence. Instructors reported respecting their students' shared expertise; despite coming from different subjects. Instructors were highly positive about the quality of resources produced by students and the creativity they brought to the module. The outcomes *"exceeded all expectations"* of the institution, producing comprehensive lecture notes, exam questions, example questions, and laboratory projects. The subsequent module was expanded after its first delivery due to the extremely popular feedback it received from student learners. This example of excellent practice featured remuneration for student partners, a combined whole-cohort and selective student approach, crediting students as co-authors on the paper reporting the project, and co-creation on a curriculum design level of high student influence. This example of excellent practice featured co-creation involving a high level of student influence on curriculum design, remuneration for student partners, a combined whole-cohort and selective student approach, and crediting students as co-authors on the paper reporting the project.

Approaches to Team Working and Collaboration in Co-creative Teams

Co-creation comes with its own strengths and challenges. Careful consideration is required when working in 'mixed' teams of instructors of various levels of seniority and students. Some particularly useful insight is given by a pair of studies by Martens *et al.*, which explore instructors' and students' perceptions of student-instructor partnerships (60, 68). Students felt they had unique insights to offer instructors and could push instructors through *"ruts"* they may be stuck in, repeating the same teaching strategies which were not fully effective. Students listed the greatest obstacle to partnership occurred when instructors were not willing to listen to their input (68).

Some instructors expressed concerns about the quality of students' inputs as they will likely have much less experience in teaching and designing curricula (60). However, student feedback was still welcomed to influence future pedagogies and refine curricula. Co-creation goes a step beyond student feedback into reciprocity, and with it, a range of benefits (1). However, where instructors are not willing to partner fully with students, co-creation may just lead to students developing voice fatigue: the sense that they are repeating themselves and not being heard (64).

Another significant lesson from these articles is the importance of clear definitions of instructors' and students' roles during co-creation. As mentioned earlier, students are likely to be inexperienced with curricula creation, and hence nebulous responsibilities may cause students to feel uncomfortable and out-of-depth, and therefore not able to contribute well. Equally, it is important instructors retain some ownership of their curricula, to prevent feeling "*out of control*" (60), and as overall responsibility for curricular quality still lies with instructors. Clear communication on the responsibilities and expectations of all parties makes for better work in a mixed team (68).

Instructors highlighted the need for a dialogue between themselves and the students (60). Opening this dialogue can be difficult in a mixed team with a power imbalance between instructors and students (63). Students may not criticize curriculum design or participate fully in co-creation due to fears of repercussions, such as in future assessments (68). Instructors may be dismissive of students' inputs due to differences in seniority and preconceived notions about the relevance of students' voices (60). Methods to encourage instructors to engage in constructive partnerships with students and open a dialogue are discussed later in the Recommendations for Future Practice section.

Team Working in an Online World

During traditional face-to-face delivery, instructors receive direct live student feedback, for example the energy and attention present in the room. The extent of this feedback varies depending on group size, classroom space, and dynamics. Online education requires instructors to make use of a somewhat different set of synchronous and asynchronous feedback mechanisms, to engage students in active learning and collect course evaluations. Co-creation is important in this online space, as this allows students to tailor the online course content to their learning needs and address gaps in understanding in a way their instructor may not have perceived without their input (49, 62, 63).

Cook-Sather discusses the use of forums to foster pedagogical discussions between instructors and students (34). Digital media, much like in the delivery of online learning, allows the breadth of inputs to increase as the asynchronous and short-form platform encourages contributions that might otherwise be lost to logistical problems. Cook-Sather notes "*virtual forms that link differently positioned people across time and space ... equalize the power*". This is an especially important point; online learning and teaching can support the student voice rather than hindering it when freedom is given and student inputs are listened to. Similarly, these online tools can support remote collaboration, as seen in the Blau & Shamir-Inbal example below (62).

Forum spaces, often incorporated into online curricula, can be a very effective means of creating a dialogue with students on a more informal level for evaluations. Structured discussions on forums have been shown to:

- Build community (62);
- Foster student voice (63);
- Be an efficient means of responding to student queries (76);
- Remove barriers to student engagement (62).

Forums are particularly useful for discussions where responses are timely, and discussions are carried out with structured roles. De Wever *et al.* analyzed 4,770 messages from asynchronous discussion forums for a first-year instructional science course where students were assigned definitive roles and found that students enacted their assigned roles (77). Following on from this, Warren notes from a similar analysis that students acted in their roles but collaborated in surprising ways, such as frequent non-critical questions from students assigned a "*Devil's advocate*" role (78). Students brought an unknown element to the course content due to their diverse backgrounds and different experiences. These examples from instructor designed and led courses could be applied to the use of forums in co-creation projects.

Blau & Shamir-Inbal write: *"The equalization effect of the digital environment, which diminishes status cues changed the power dynamic, promoted students' active participation and their pedagogical partnership with the instructor"* (62). Their paper describes the co-creation of a module in the digital space, involving customization of learning by 54 distance learning education graduate students. In addition to forums, students were given editing access to course materials through Google Apps for Education. This fostered collaboration with instructors in editing, restructuring, and collating relevant course materials. The authors found that the combination of several digital platforms created an online learning community that removed many of the differences between students and instructors. This approach to online learning could be applied to a self-directed chemistry course, where students are given participatory freedom to edit course guides as they learn.

Building Student Influence and Power Slowly: Easing into Co-creation

Bovill argues for pragmatism:

"It is unrealistic to assume that many academic staff and students will change to co-creating entire programmes without first perhaps experimenting on a smaller scale" (58).

We recommend beginning with a smaller project to give students and instructors this experience with the process of co-creation, building up to deliver on a larger project.

Little adopted a comprehensive style of co-creation for curriculum design that includes many levels of participation (61). A student ambassador network was initiated by the faculty, and as time went on the students set their own agenda transforming into decision-makers rather than adopting advisory type roles similar to student body representatives. Students and instructors developed and wrote up a partnership strategy document outlining the roles and responsibilities of several working groups, and this was determined by Martens *et al.* to be an important example of good practice (60). These groups were given a small budget for developing curricula and hence student ambassadors were paid for their work. Students worked with specific instructors in their own departments to negotiate a curriculum but also fed back to the broader network about their processes of curriculum negotiation.

Although students reported being nervous at first to approach instructors, an *"upward spiral"* of confidence was reported over the course of the project. Students began to voice their opinions during negotiations more after a working relationship was developed with their corresponding staff member. Hence, the mixed team became more effective over time. In addition, the student ambassadors network gave a place for students to exchange their experiences and frustrations outside of their individual departments, contributing to the improvements in mixed teamwork.

Impact of Partnership on Engagement and Experience in Learning and Teaching

Partnership in the development of learning and teaching leads to benefits for both instructors and students involved (79, 80). Beyond the collaborators involved, some of the benefits of co-creative partnerships extend to students not directly involved, such as increased instructional quality, increased quality and relevance of teaching materials, and increased student-instructor contact (enabled through freeing up staff time to engage more students). Wider adoption of student partnerships at the institution level has the potential to transform institutional cultures with consequences for everyone (instructors and students), not just those directly involved in co-creation projects. Potential negative outcomes of partnership such as challenges maintaining the quality control of outputs, would also likely impact the whole cohort regardless of participation with pedagogical partnership.

Experience of Student Partners

Through participation in collaborative pedagogical partnerships with faculty, students develop knowledge, skills, and agency (*81*). Amongst the skills developed are many valued graduate attributes including verbal and written communication and effective team-working skills (*80*). The collaboration encourages discussions between student and faculty partners which clarify learning expectations and pedagogic rationales; this, in turn, improves student learning outcomes (*82*). This approach can be incorporated directly as part of the learning and teaching for a module, where students create materials for modules they are currently studying, as shown earlier by Hubbard *et al.* with student-created digital organic chemistry tutorials (*51*).

Bovill *et al.* investigated student-partnerships using case study methodology (*83*), they reported that working in partnership with instructors can benefit students' motivation, confidence, commitment, and perception of shared responsibility for learning (*84*). By working in partnership with students to influence learning and teaching, the curriculum becomes more relevant to students' lives and thereby more engaging to students. Engagement is linked to learning gain and attainment and hence increased student engagement fosters further benefits in learning and achievement (*85*).

Furthermore, taking part in extra-curricular activities such as collaborative partnerships can increase students' sense of belonging which is closely associated with wellbeing, academic attainment, and retention (*86*). However, it is important to note that performative exercises where the students do not feel listened to or valued can lead to feelings of alienation and voicelessness (*87*).

Experience of Instructors

Instructors benefit and learn from taking part in collaborative partnerships with students too. Autobiographical reflections and analysis of case studies has highlighted that faculty partners see benefits in terms of their engagement with and satisfaction with scholarship and teaching (*79, 84*). The incorporation of multiple perspectives can uncover alternative conceptions and preconceptions allowing for assumptions to be questioned and constructing new and deeper understanding into pedagogy, discipline-specific knowledge, and student voices (*88*).

Engaging in collaborative partnerships with students can encourage academic development through the scholarship of learning and teaching. Instructors further develop their understanding of pedagogy through explaining these concepts to student partners and discussing the design of curricula. These student-instructor discussions may encourage further scholarship through the review of pedagogical literature. Barnett and Hallam argue that partnership work can counter the tensions which can exist between teaching and research by dissolving the barriers between the two (*89*).

Relationship Building

"When students are listened to by academic staff, both the students and the staff see themselves differently." Bovill, Cook-Sather, and Felten, 2011 (*84*)

Partnership challenges traditional hierarchies and roles. Engaging in student partnerships can alter conceptions of teaching and learning, reframing the learning process as *"a collaborative venture of students and teacher"* which has consequences for the relationship between student and instructor (*84*). Cook-Sather reflected that partnerships appear *"to inspire greater openness to, and appreciation of, differences and to foster deeper connection and empathy across student and staff positions, perspectives and cultural identities"* (*90*). A shared commitment to learning and teaching was described by Davis and Sumara as *"the power of positioning students as co-creators of learning"* (*91*). Fostering productive and meaningful social interactions between students and instructors is particularly important for remote and online learning productivity. These social factors positively affect the learning and teaching experience of both instructors and students, encourage a sense of community, and improve educational outcomes for students (*92*).

Spotlight on Equality, Diversity, Inclusion, and Respect: Can Adoption of Co-creation Pedagogies further EDIR Goals?

"Substantial benefits can arise from viewing diverse and often excluded students as valuable co-researchers, consultants and pedagogical co-designers." Bovill et al. 2016 (*90*)

Inequalities in Participation, Experience, and Attainment in STEM Higher Education

Reviewing the statistics for participation in academia and STEM careers it is clear that many areas of diversity are underrepresented and underserved including women, people of color, and people with disabilities (*93–97*). Approximately 1 in 5 people in the UK and 1 in 8 people in the US identify as having a disability (*98*), however, the declaration rate is lower for STEM students, and lower still for STEM academics (*95*). UK higher education statistics show a considerable gap between the proportion of white British students receiving first or 2:1 degree classifications compared to UK-domiciled students from minority ethnic groups; with the greatest gap observed between white British students and black British students (in 2015/16 the numbers were 78.8% and 50.5% respectively) (*99*, *100*).

The individual experiences of minoritized students are diverse and intersecting with other identities, an individual may face multiple biases simultaneously (*101–103*). Existing inequalities in education and wider society have been exacerbated by the Covid-19 pandemic as highlighted in the UN Sustainable Development Goals Report, 2020 (*104*). These inequalities must be urgently addressed. A recent editorial in the Journal of Chemical Education called for the chemistry community to *"adopt practices and research scholarship that respond to the needs of diversity, equity, inclusion, and respect"* (*105*).

A major multi-disciplinary study of UK higher education institutions reported that *"The lack of a sufficiently diverse or decolonised curriculum and faculty meant it was often difficult for black students to be able to connect content and assessments directly to their own lived realities"* (*106*). Inequalities in society also manifest in unequal access to and participation in faculty-student partnerships (*65*). This has major consequences on which student voices are heard and incorporated in curriculum development and design. In light of this, many authors have cautioned against homogenizing the student voice, instead advising higher education institutions to recognize students' nuanced and individual experiences (*107–111*).

Inclusive Online Learning and Teaching

Inequities in access to the benefits of technology in society are commonly referred to as The Digital Divide (*112*). In education, these inequities may be caused by a lack of access to suitable electronic devices, poor internet connectivity (WIFI/data), and/or undeveloped digital competencies required to effectively use digital resources (*113, 114*). Furthermore, variable levels of digital accessibility creates barriers that make learning online more challenging for disabled students/students with disabilities (*115*).

Online Learning must provide an equitable experience for all students and consider individual access needs in an inclusive manner (*116, 117*). Dewsbury defines inclusive pedagogy as "*a philosophy of teaching that provides equal opportunities for all students to have a successful learning experience*" (*118*). Factors affecting a student's achievement and feelings of inclusion are: a sense of belonging, self-efficacy, identity, and stereotype threat (*119*). Earlier, we established that co-creation pedagogies increase student engagement, and this is a key predictor of academic success in higher education (*120*).

Cook-Sather identified three key ways in which student partnerships promote inclusivity: positioning students to bring their identities and experiences to influence developing inclusive classrooms, drawing on their experiences to recommend pedagogical approaches that are responsive to a greater diversity of students, and by making faculty aware of pedagogical practices they already use which foster inclusivity (*121, 122*). Practical considerations to ensure partnerships are inclusive and equitable are given at the end of this section.

Student Partnerships Furthering Social Justice in Higher Education

> "*The promise of giving voice to under-represented and marginalized groups has been a mainstay of emancipatory agendas in educational research.*" Julie McLeod, 2011 (*109*)

Collaborative pedagogies, including Students as Partners approaches, can be tools to further social justice, disrupting the *status quo*, encouraging equity, and raising student voice (*109*). This is unlikely to happen by accident. Ensuring that all students can participate in collaborative partnerships and actively including minoritized students leads to the inclusion of diverse perspectives into the scholarship of learning and teaching and hence the diversity and relevance of the curricula (*123*). Engaging with diverse partnerships encourage faculty to reflect on and challenge conscious and unconscious biases which influence what needs we perceive our students to have (*124*). In addition to improving the experience of cohorts, valuing the diverse lived experiences of students, is viewed positively by the student partners too. Delgado-Bernal wrote that "*student consultants report that their experiences and knowledge are viewed as resources rather than deficits*" (*125*). Collaborative partnerships can build trust, empower, and engage with students from minoritized groups (*123*). However, Cook-Sather warns that students also possess biases and privileges which can exacerbate discrimination of instructors with minoritized identities (*64*), which has been evident in studies of student evaluation tools in higher education (*126*). With careful consideration of equity, collaboration can foster mutual respect and develop instructors' voices concerning their own EDIR issues, as demonstrated by Perez (*127*).

White *et al.* recommend developing a diverse curriculum that contains voices of scientists with different social identities and cultural backgrounds, and developing learning environments and communities where minoritized students feel welcome, valued, and validated (*128*). Diverse curricula are educationally effective, helping to foster feelings of belonging, promote academic skill development and encourage deeper learning (*129, 130*). Further evidence for the benefits of diverse curricula is illustrated by in a study by Good *et al.* evaluating the effect of a novel online STEM course featuring multicultural or "*colorblind*" (*131*) content ($N = 688$), students of color showed greatest STEM performance when studying the multicultural curriculum (*132*). The STEM courses included math, physics, and chemistry lessons and were delivered to first-year college students. Interestingly, although white students' STEM performance was not affected by changing the curriculum, the multicultural instructor was perceived by white students to be most biased, whereas the multicultural instructor was perceived by "*students of color*" to be the least biased.

Students with disabilities face additional barriers in equitable access to online learning when the content or tools required are not fully accessible. In STEM scientific and mathematical notation can be inaccessible to screen readers (*97*). Students with lived experience of disability can provide invaluable feedback on the user experience and draw on their lived experience to develop educational materials which are born-accessible. By incorporating their ideas and insights into the design of courses, the need to make adaptions retroactively may be reduced, however, every student is individual, and it cannot be assumed that the lived experience of students with disability will be similar.

Examples from Practice: Furthering EDIR Goals through Student Partnership in Chemistry Higher Education

There are currently few published examples of studies in chemistry education where student partnerships have been adopted to further EDIR goals. We explore relevant examples below.

Student partners co-developed a novel sustainable chemistry module where they adopted a systems-thinking approach to explore sustainable futures using the United Nations Sustainable Development Goals (*133*) as a framework (*134*). In this example of partnership, the collaborative team developed a course curriculum that explored equality and justice aspects of sustainability. The team reflected that collaboration with students ensured that delivery of the course would address student concerns (such as timetabling and workload).

Goethe and Colina empowered chemistry and chemical engineering students in their classes from minoritized groups to "*enrich the classroom*", and felt that in doing so both educational outcomes and retention were improved (*135*). The authors encourage teaching faculty to incorporate diversity into their instruction, through which we can "*foster a more intellectually challenging learning environment in which all students, and thus society, will benefit*". How to do this? Goethe and Colina recommended encouraging discussions in the classroom about diversity of perspective and experience, using discussion of disciplinary perspectives as a scaffold to introduce discourse on diverse identities, adopting culturally relevant pedagogies (*136*), and raising the voices of diverse perspectives within our classrooms.

Dunlop *et al.* reported a further example of encouraging discussion and debate between students with different disciplinary perspectives (*137*). An interdisciplinary team of undergraduate student partners (from the Departments of Chemistry, Education, and Philosophy) worked together to develop and deliver outreach activities for high school aged students exploring philosophical questions in chemistry contexts. The experience of the student partners was evaluated using a capabilities approach, encompassing wellbeing and abilities, rather than student satisfaction. They reported that this extracurricular activity created a space for students to "*build capabilities to flourish through philosophical dialogue about chemistry*". The authors argued that philosophical dialogue can challenge students to think about their subject in a new way and is often missing from chemistry education and along with it, opportunities for discomfort that can contribute to students' capabilities to achieve happiness and well-being are missed. This study suggests that encouraging diverse discourse can lead to gains in terms of improved wellbeing for the whole cohort.

Student partners have also co-developed educational interventions delivered to students with similar racial heritage. LatinX (Latine) graduate students engaged in collaborative partnerships with faculty to devise, implement, and assess the impact of interventions intended to promote active learning in classrooms at two Hispanic serving institutions (higher education institutions with enrolment of undergraduate full-time equivalent students that is at least 25% Hispanic students) (*138*). Graduate Student Partners were provided training on restructuring college courses via backwards design before collaboration to support equitable partnership with faculty partners.

Recommendations for Future Practice

Key examples of good practice in co-creation distilled from the literature are collated in figure 3 below, sorted by topic.

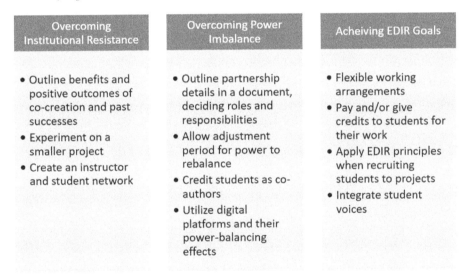

Figure 3. Examples of key good practices in co-creation in higher education. [Transcription left to right: Overcoming Institutional Resistance: outline benefits and positive outcomes of co-creation and past success, experiment on a smaller project, create an instructor and student network. Overcoming Power Imbalance: outline partnership details in a document, deciding roles and responsibilities, allow adjustment period for power to rebalance, credit students as co-authors, utilize digital platforms and their power-balancing effects. Achieving EDIR Goals: flexible working arrangements, Pay and/or giving credits to students for their work, Apply EDIR principles when recruiting students to projects, Integrate student voices. End transcription]

Including the Whole Cohort

The aim of co-creation is to improve teaching and learning by valuing students' input and collaboration with instructors. This cannot be done without attempts to include the whole cohort of a particular course, especially given that it has been shown that some students from minoritized groups may feel co-creation "*is not for them*" (*64*). However, limited resources and the risk of over-complicating curriculum design mean high-participation co-creation with hundreds of students on a given course is not feasible.

Bovill makes the case for including the whole cohort on the basis that in reported literature whole-cohort approaches strengthened the relationship between students and instructors **more** than select approaches (*139*). Bovill suggests this is due to the increased inclusion and a greater proportion of students feeling that their voice was heard. The authors recommend that methods of co-creation are considered carefully to appeal to students of different backgrounds and provide multiple ways of being involved, such that students can choose a way that suits their strengths and time limitations.

In online chemical education, we recommend a mixed approach with two phases to be taken. Online delivery offers unique opportunities for students to personalize and adapt their learning to suit their needs and circumstances. A mixed approach was adopted by the University of Glasgow, where a small number of students were selected to collaborate on weekend or summer projects developing curricula from the ground up with instructors, whilst the whole cohort was given the opportunity to audit, evaluate, customize and influence the teaching during implementation (*75*). This might take the form of creating digital resources as assignments that then become teaching resources, customizable assessment formats, and course content that is presented in several ways so a student can customize how they learn. Students and instructors can be surveyed as to course needs and subsequently small mixed teams can develop curricula and materials to address the needs and solve problems highlighted by the surveying.

Addressing the Balance of Power

Student partnerships have the power to break down traditional hierarchies, enabling and raising the student voice.

Care must be taken to safeguard students in these mixed teams. The power imbalance must be addressed directly, and students **listened to**. Students involved in co-creation are usually current students in the department they are creating for— what steps are taken to ensure that they feel comfortable to make suggestions to an instructor who may be involved in the assessment of their degree? Are students' contributions and suggestions respected? Are students credited and paid for their work on curricula?

Bovill et al. recommend an active recognition that traditional student and instructor roles and hierarchies are socially constructed and changeable (*90*). This opens up the discussion to new ways of considering teaching and learning. The focus is shifted from constraints and issues with co-creation to potential opportunities. The authors also recommend clearly defining the roles and responsibilities of students and instructors, and potential benefits and challenges. This gives appropriate agency to students and instructors and can encourage institutional support for a co-creation project by highlighting positive outcomes (*90*). As discussed above, digital tools such as discussion forums can often disrupt the balance of power allowing for more effective communication within the team.

To increase student voice, students can also be encouraged to contribute to curriculum theory: contribution **of** curricula as opposed to **in** curricula. This will maximize the application of their expertise as learners and embed co-creation in the pedagogy of an institution. This can be done by extending current discussions of curriculum theory to include students, through the use of boards, mixed networks, and student consultants on course design (*59*).

Evaluating and Collecting Evidence

There are relatively few published examples of co-creative approaches in chemistry in higher education, but there have likely been many more cases. Reflecting on, evaluating, and disseminating co-creation projects will share learned wisdom and 'normalize' co-creation, encouraging further adoption. Woolmer's thesis includes examples of interview questions used to explore experiences of co-creation which make for an excellent starting point (*140*). Additionally, it would be interesting to explore the experiences of cohorts using co-created materials or taking co-created modules but who were not involved in their development. Consequences of co-creation pedagogies for these groups of students has been largely unexplored so far.

Authorship, Recognition, and Remuneration

Mercer-Mapstone *et al.* note that "*although students and staff may work collaboratively on a project, this does not always result in co-authorship*" (*53*). Only a third (N=65) of papers reviewed had students listed as co-authors and 88% listed instructors as the first author. This is not in keeping with the co-creation principles of respecting and valuing students' inputs.

Principles of ethical authorship apply in co-creation just as they would in a mixed research group (*49*). Although instructors may take the lead on quality assurance and final delivery of a curriculum, students' contributions must be respected, credited, and recognized just as an undergraduate student performing laboratory work must be credited in a research article. In some cases, such as input deriving from an invisible disability or marginalized identity, students may wish to remain uncredited, in which case their wishes should be respected. Authorship proves a particular challenge in online education, where many tasks and resources are scattered and remixed. Simple steps such as a 'creators' biography' section or page on online portals may help.

The case studies discussed by Bovill raise an example of good practice: remunerating students for their work, which is necessary as a sign of respect for their input (*58*). Students should be given remuneration, or equivalent course credits for the time spent working on a co-creation project. This respects the student as a partner in creation and removes financial barriers to participation (*58*). However, many instructors wishing to engage in pedagogical partnerships find it challenging to obtain funding for scholarship of teaching and learning projects. In some cases, it may be possible to access funding from professional bodies and charities but ultimately institutions must invest in supporting inclusive partnerships if they are to be sustainable. Time restrictions help to safeguard students, who should not be asked to input many hours alongside their studies in the semester, already a full-time undertaking. Some issues can arise with the availability of academic staff over the summer, typically a time for academic research, so flexibility may be required.

Employ Inclusive Recruitment and Working Practices

Recruitment of Student Partners

Whether in addition to whole cohort approaches or when it is not possible or practical to consult the whole cohort, it is vital to ensure the recruitment of student partners is fair and equitable. Preconceptions of students based on characteristics such as race and together with unconscious biases could lead to prejudicial and discriminatory recruitment practices. Felten *et al.* warned that that common methods for selecting student co-researchers such as open enrolment into classes or direct invitation from a faculty member can reduce the diversity of student partnerships (*65*). Active recruitment of underrepresented/minoritized students and valuing diverse lived experiences when shortlisting can help to ensure diverse perspectives are included in collaborations. This is particularly important when developing interventions that will be delivered to minoritized students.

Example of active recruitment of student partners from underrepresented backgrounds: The Students as Learners and Teachers program at Bryn Mawr College in Pennsylvania selects underrepresented students on campus to serve as pedagogic consultants to faculty members (*90*). Students were invited who had participated previously in diversity initiatives or were taking courses on multicultural education and the students were paid by the hour for their participation. The students identified what the faculty members were already doing to create classrooms that were welcoming to a diversity of students and suggested additional actions those faculty members could do to make their classrooms more welcoming.

Listening to and Valuing Student Partner Ideas and Contributions

Students should be valued as peers and their expertise in their own learning and lived experiences should be recognized. Seale *et al.* warn that if we "*ignore issues of power and resistance, we will fall far short of the vision of student engagement and the ideals of strong participation and expression of student voice*" (*141*).

Inclusive Working Practices

Inclusive working practices such as part-time working, flexibility, or remote working can increase the number and diversity of students who can participate in pedagogical partnerships. For example, the ability to work from home flexibly can remove barriers for disabled students, allowing them to organize their work in a way that suits the individual and reduce commuting (*95*). Reducing commuting also reduces the costs and time involved in taking part in collaborations.

Training for Pedagogical Success

Providing students with training has obvious benefits in terms of their experience and skills development but it can also help to ensure effective partnership, empowering students' voices, and ameliorating faculty fears around student experience and skills and the quality of student-generated content. Clearly outlining the roles and expectations of students and instructors, as well as the structure of a co-creation project, can also assuage fears about experience gaps (*60*).

Conclusion

It is vital to engage our student communities with effective, equitable, and enjoyable online learning. The adoption of collaborative pedagogies with students as partners in the design of teaching and learning is an effective framework to achieve high-quality learning both in the classroom and online. Partnership approaches are inherently more efficient and inclusive as they deliver curricula and course materials that are born student-voice-informed.

In this chapter, we have explored different levels of student involvement, approaches, and strategies for effective student-instructor collaboration, including ethics, power distribution, credit, responsibility, and dissemination. Key benefits were identified including skills development for students and instructors, increase in student engagement and agency, and the development of learning innovations that meet real rather than perceived student needs. Students are not a homogenous group and student partnership can allow for the inclusion of diverse student voices, leading to improvements in the equity of teaching effectiveness and engagement, and relevance of the curriculum. However, without careful consideration of the barriers which can prevent traditionally minoritized students from taking part, inequitable participation can skew the 'student voice' towards relatively privileged students and exacerbate existing inequalities.

In many cases, the rapid pivot to online learning necessitated by the Covid-19 pandemic did not allow time for collaborative approaches. Looking ahead, we recommend adopting democratic approaches, with students as partners in the development of online teaching curricula and materials. We anticipate an increase in the provision of online learning and teaching as the higher education sector reflects and moves on from the Covid-19 pandemic. Many are questioning the place of the traditional live lecture and considering adopting more blended and flipped teaching pedagogies. Student partners should play a pivotal role in the design, delivery, and evaluation of these new online chemistry curricula.

References

1. Bovill, C.; Cook-Sather, A.; Felten, P. *Engaging Students as Partners in Learning and Teaching: A Guide for Faculty*; Jossey-Bass: 2014.
2. Könings, K. D.; Mordang, S.; Smeenk, F.; Stassen, L.; Ramani, S. Learner involvement in the co-creation of teaching and learning: AMEE Guide No. 138. *Med. Teach.* **2020**, 1–13.
3. Johnstone, A. H. Why is science difficult to learn? Things are seldom what they seem. *J. Comput. Assist. Learn.* **1991**, *7*, 75–83.
4. Taber, K. S. Revisiting the chemistry triplet: drawing upon the nature of chemical knowledge and the psychology of learning to inform chemistry education. *Chem. Educ. Res. Pract.* **2013**, *14*, 156–168.
5. Sweller, J. Cognitive load during problem solving: Effects on learning. *Cogn. Sci.* **1988**, *12*, 257–285.
6. Sweller, J.; van Merrienboer, J. J. G.; Paas, F. G. W. C. Cognitive Architecture and Instructional Design. *Educ. Psychol. Rev.* **1998**, *10*, 251–296.
7. Martens, S. E.; Meeuwissen, S. N. E.; Dolmans, D. H. J. M.; Bovill, C.; Könings, K. D. Student participation in the design of learning and teaching: Disentangling the terminology and approaches. *Med. Teach.* **2019**, *41*, 1203–1205.
8. *The Journal of Educational Innovation, Partnership and Change.*
9. *International Journal for Students as Partners.*
10. Bovill, C.; Bulley, C. J. A model of active student participation in curriculum design: Exploring desirability and possibility. In *Improving Student Learning (18) Global theories and local practices: Institutional, disciplinary and cultural variations*; Rust, C., Ed.; The Oxford Centre for Staff and Educational Development: 2011; pp 176–188.
11. Cooper, M. M.; Stowe, R. L. Chemistry Education Research—From Personal Empiricism to Evidence, Theory, and Informed Practice. *Chem. Rev.* **2018**, *118*, 6053–6087.
12. Seery, M. K.; O'Connor, C., E-Learning and Blended Learning in Chemistry Education. In *Chemistry Education: Best Practices, Opportunities and Trends*; Wiley: 2015; pp 651–670.
13. World Health Organization. *Coronavirus Disease (COVID-19) Pandemic.* https://www.who.int/health-topics/coronavirus#tab=tab_1 (date accessed May 2020).
14. Villanueva, M. E.; Camilli, E.; Chirillano, A. C.; Cufré, J. A.; De Landeta, M. C.; Rigacci, L. N.; Velazco, V. M.; Pighin, A. F. Teaching instrumental analytical chemistry during covid-19

times in a developing country: Asynchronous versus synchronous communication. *J. Chem. Educ.* **2020**, *97*, 2719–2722.

15. Parsons, A. F. Exploring Everyday Chemistry: The Effectiveness of an Organic Chemistry Massive Open Online Course as an Education and Outreach Tool. *J. Chem. Educ.* **2020**, *97*, 1266–1271.

16. Blackburn, R. A. R.; Villa-Marcos, B.; Williams, D. P. Preparing Students for Practical Sessions Using Laboratory Simulation Software. *J. Chem. Educ.* **2019**, *96*, 153–158.

17. Jones, E. V.; Shepler, C. G.; Evans, M. J. Synchronous Online-Delivery: A Novel Approach to Online Lab Instruction. *J. Chem. Educ.* **2021**, *98*, 850–857.

18. Mahaffey, A. L. 1–2–3 Benchtop to Laptop: Teamwork of an Educator and Instructional Designer to Convert a Popular Ksp and Titration Lab to an Online Module. *J. Chem. Educ.* **2021**, *98*, 1928–1936.

19. Woelk, K.; Whitefield, P. D. As Close as It Might Get to the Real Lab Experience—Live-Streamed Laboratory Activities. *J. Chem. Educ.* **2020**, *97*, 2996–3001.

20. Irby, S. M.; Borda, E. J.; Haupt, J. Effects of Implementing a Hybrid Wet Lab and Online Module Lab Curriculum into a General Chemistry Course: Impacts on Student Performance and Engagement with the Chemistry Triplet. *J. Chem. Educ.* **2018**, *95*, 224–232.

21. Tee, N. Y. K.; Gan, H. S.; Li, J.; Cheong, B. H.-P.; Tan, H. Y.; Liew, O. W.; Ng, T. W. Developing and Demonstrating an Augmented Reality Colorimetric Titration Tool. *J. Chem. Educ.* **2018**, *95*, 393–399.

22. Müssig, J.; Clark, A.; Hoermann, S.; Loporcaro, G.; Loporcaro, C.; Huber, T. Imparting Materials Science Knowledge in the Field of the Crystal Structure of Metals in Times of Online Teaching: A Novel Online Laboratory Teaching Concept with an Augmented Reality Application. *J. Chem. Educ.* **2020**, *97*, 2643–2650.

23. Norrby, M.; Grebner, C.; Eriksson, J.; Boström, J. Molecular Rift: Virtual Reality for Drug Designers. *J. Chem. Inf. Model.* **2015**, *55*, 2475–2484.

24. Ferrell, J. B.; Campbell, J. P.; McCarthy, D. R.; McKay, K. T.; Hensinger, M.; Srinivasan, R.; Zhao, X.; Wurthmann, A.; Li, J.; Schneebeli, S. T. Chemical Exploration with Virtual Reality in Organic Teaching Laboratories. *J. Chem. Educ.* **2019**, *96*, 1961–1966.

25. Rychkova, A.; Korotkikh, A.; Mironov, A.; Smolin, A.; Maksimenko, N.; Kurushkin, M. Orbital Battleship: A Multiplayer Guessing Game in Immersive Virtual Reality. *J. Chem. Educ.* **2020**, *97*, 4184–4188.

26. Dunnagan, C. L.; Gallardo-Williams, M. T. Overcoming Physical Separation During COVID-19 Using Virtual Reality in Organic Chemistry Laboratories. *J. Chem. Educ.* **2020**, *97*, 3060–3063.

27. Lee, M. W. Online Teaching of Chemistry during the Period of COVID-19: Experience at a National University in Korea. *J. Chem. Educ.* **2020**, *97*, 2834–2838.

28. Sarju, J. P. Rapid Adaptation of a Traditional Introductory Lecture Course on Catalysis into Content for Remote Delivery Online in Response to Global Pandemic. *J. Chem. Educ.* **2020**, *97*, 2590–2597.

29. Wild, D. A.; Yeung, A.; Loedolff, M.; Spagnoli, D. Lessons Learned by Converting a First-Year Physical Chemistry Unit into an Online Course in 2 Weeks. *J. Chem. Educ.* **2020**, *97*, 2389–2392.

30. Rodríguez Núñez, J.; Leeuwner, J. Changing Courses in Midstream: COVID-19 and the Transition to Online Delivery in Two Undergraduate Chemistry Courses. *J. Chem. Educ.* **2020**, *97*, 2819–2824.

31. Seery, M. K. Flipped learning in higher education chemistry: emerging trends and potential directions. *Chem. Educ. Res. Pract.* **2015**, *16*, 758–768.

32. Wikandari, R.; Putro, A. W.; Suroto, D. A.; Purwandari, F. A.; Setyaningsih, W. Combining a Flipped Learning Approach and an Animated Video to Improve First-Year Undergraduate Students' Understanding of Electron Transport Chains in a Biochemistry Course. *J. Chem. Educ.* **2021**, *98*, 2236–2242.

33. Turner, K. L.; Chung, H. Transition to eBook Provision: A Commentary on the Preferences and Adoption of eBooks by Chemistry Undergraduates. *J. Chem. Educ.* **2020**, *97*, 1221–1225.

34. Grinias, J. P.; Smith, T. I. Preliminary Evidence on the Effect of an Open-Source Textbook in Second-Year Undergraduate Analytical Chemistry Courses. *J. Chem. Educ.* **2020**, *97*, 2347–2350.

35. Sansom, R. L.; Clinton-Lisell, V.; Fischer, L. Let Students Choose: Examining the Impact of Open Educational Resources on Performance in General Chemistry. *J. Chem. Educ.* **2021**, *98*, 745–755.

36. Poole, M. J.; Glaser, R. E. Organic Chemistry Online: Building Collaborative Learning Communities through Electronic Communication Tools. *J. Chem. Educ.* **1999**, *76*, 699–703.

37. Ealy, J. B. Development and Implementation of a First-Semester Hybrid Organic Chemistry Course: Yielding Advantages for Educators and Students. *J. Chem. Educ.* **2013**, *90*, 303–307.

38. Saar, A.; McLaughlin, M.; Barlow, R.; Goetz, J.; Adediran, S. A.; Gupta, A. Incorporating Literature into an Organic Chemistry Laboratory Class: Translating Lab Activities Online and Encouraging the Development of Writing and Presentation Skills. *J. Chem. Educ.* **2020**, *97*, 3223–3229.

39. Jones, M. L. B.; Seybold, P. G. Combining Chemical Information Literacy, Communication Skills, Career Preparation, Ethics, and Peer Review in a Team-Taught Chemistry Course. *J. Chem. Educ.* **2016**, *93*, 439–443.

40. Williamson, V. M.; Zumalt, C. J. How do general chemistry students' impressions, attitudes, perceived learning, and course performance vary with the arrangement of homework questions and E-text? *Chem. Educ. Res. Pract.* **2017**, *18*, 785–797.

41. Aricò, F. R.; Lancaster, S. J. Facilitating active learning and enhancing student self-assessment skills. *Int. Rev. Econ. Educ.* **2018**, *29*, 6–13.

42. Gagan, M. *Review of the Student Learning Experience in Chemistry*; The Higher Education Academy Physical Sciences Centre: Hull, U.K., 2009.

43. Talanquer, V.; Bucat, R.; Tasker, R.; Mahaffy, P. G. Lessons from a Pandemic: Educating for Complexity, Change, Uncertainty, Vulnerability, and Resilience. *J. Chem. Educ.* **2020**, *97*, 2696–2700.

44. Seery, M. K.; McDonnell, C. The Application of Technology to Enhance Chemistry Education. *Chem. Educ. Res. Pract.* **2013**, *14*, 227–228.

45. Kiernan, N.; Manches, A.; Seery, M. K. The Role of Visuospatial Thinking in Students' Predictions of Molecular Geometry. *Chem. Educ. Res. Pract.* **2021**, *22*, 626–639.

46. Krishnan, M. S.; Brakaspathy, R.; Arunan, E. Chemical Education in India: Addressing Current Challenges and Optimizing Opportunities. *J. Chem. Educ.* **2016**, *93*, 1731–1736.
47. de Freitas, S. I.; Morgan, J.; Gibson, D. Will MOOCs transform learning and teaching in higher education? Engagement and course retention in online learning provision. *Br. J. Educ. Technol.* **2015**, *46*, 455–471.
48. Carey, P. Representation and student engagement in higher education: A reflection on the views and experiences of course representatives. *J. Furth. High. Educ.* **2013**, *37*, 71–88.
49. McCollum, B.; Morsch, L.; Pinder, C.; Ripley, I.; Skagen, D.; Wentzel, M. Multi-dimensional trust between partners for international online collaborative learning in the Third Space. *IJSaP* **2019**, *3*, 50–59.
50. Santos-Díaz, S.; Towns, M. H. Chemistry outreach as a community of practice: investigating the relationship between student-facilitators' experiences and boundary processes in a student-run organization. *Chem. Educ. Res. Pract.* **2020**, *21*, 1095–1109.
51. Hubbard, B. A.; Jones, G. C.; Gallardo-Williams, M. T. Student-Generated Digital Tutorials in an Introductory Organic Chemistry Course. *J. Chem. Educ.* **2019**, *96*, 597–600.
52. Gupta, N.; Nikles, J. Impact of Student-Created Mechanism Videos in Organic Chemistry 2 Labs. *ACS Symp. Ser.* **2019**, *1325*, 107–117.
53. Mercer-Mapstone, L.; Dvorakova, S. L.; Matthews, K. E.; Abbot, S.; Cheng, B.; Felten, P.; Knorr, K.; Marquis, E.; Shammas, R.; Swaim, K. A Systematic Literature Review of Students as Partners in Higher Education. *IJSaP* **2017**, *1*.
54. Davis, C.; Parmenter, L. Student-staff partnerships at work: epistemic confidence, research-engaged teaching and vocational learning in the transition to higher education. *Educ. Action Res.* **2020**, 1–18.
55. Rainey, K.; Dancy, M.; Mickelson, R.; Stearns, E.; Moller, S. Race and gender differences in how sense of belonging influences decisions to major in STEM. *Int. J. STEM Educ.* **2018**, *5*, 10.
56. Master, A.; Meltzoff, A. N. Cultural stereotypes and sense of belonging contribute to gender gaps in STEM. *Int. J. Gend., Sci. Tech.* **2020**, *12*, 152–198.
57. Fink, A.; Frey, R. F.; Solomon, E. D. Belonging in general chemistry predicts first-year undergraduates' performance and attrition. *Chem. Educ. Res. Pract.* **2020**, *21*, 1042–1062.
58. Bovill, C. An investigation of co-created curricula within higher education in the UK, Ireland and the USA. *Innov. Educ. Teach. Int.* **2014**, *51*, 15–25.
59. Bovill, C.; Woolmer, C. How conceptualisations of curriculum in higher education influence student-staff co-creation in and of the curriculum. *High. Educ.* **2018**, *78*, 407–422.
60. Martens, S. E.; Wolfhagen, I. H. A. P.; Whittingham, J. R. D.; Dolmans, D. H. J. M. Mind the gap: Teachers' conceptions of student-staff partnership and its potential to enhance educational quality. *Med. Teach.* **2020**, *42*, 529–535.
61. Little, S. Promoting a collective conscience: designing a resilient staff–student partnership model for educational development. *Int. J. Acad. Dev.* **2016**, *21*, 273–285.
62. Blau, I.; Shamir-Inbal, T. Digital technologies for promoting "student voice" and co-creating learning experience in an academic course. *Instr. Sci.* **2018**, *46*, 315–336.
63. Cook-Sather, A. Virtual forms, actual effects: how amplifying student voice through digital media promotes reflective practice and positions students as pedagogical partners to

63. prospective high school and practicing college teachers. *Br. J. Educ. Technol.* **2017**, *48*, 1143–1152.
64. Cook-Sather, A. Respecting voices: how the co-creation of teaching and learning can support academic staff, underrepresented students, and equitable practices. *High. Educ.* **2020**, *79*, 885–901.
65. Felten, P.; Bagg, J.; Bumbry, M.; Hill, J.; Hornsby, K.; Pratt, M.; Weller, S. A Call for Expanding Inclusive Student Engagement in SoTL. *Teach. Learn. Inq.* **2013**, *1*, 63–74.
66. Flint, A. Moving from the fringe to the mainstream: opportunities for embedding student engagement through partnership. *Student Engagement in Higher Education* **2016**, *1*.
67. Cook-Sather, A. Dialogue across differences of position, perspective, and identity: Reflective practice in/on a student-faculty pedagogical partnership program. *Teach. Coll. Rec.* **2015**, *117*.
68. Martens, S. E.; Spruijt, A.; Wolfhagen, I. H. A. P.; Whittingham, J. R. D.; Dolmans, D. H. J. M. A students' take on student–staff partnerships: experiences and preferences. *Assess. Eval. High. Educ.* **2019**, *44*, 910–919.
69. Supalo, C. A. In *ConfChem Conference on Interactive Visualizations for Chemistry Teaching and Learning: Concerns Regarding Accessible Interfaces for Students Who Are Blind or Have Low Vision*; American Chemical Society: 2016; pp 1156–1159.
70. Supalo, C. Techniques to enhance instructors' teaching effectiveness with chemistry students who are blind or visually impaired. *J. Chem. Educ.* **2005**, *82*, 1513–1518.
71. Morsch, L. A. Student Authored Video Vignettes in Chemistry. *E-Mentor* **2017**, *3*, 25–32.
72. Jordan, J. T.; Box, M. C.; Eguren, K. E.; Parker, T. A.; Saraldi-Gallardo, V. M.; Wolfe, M. I.; Gallardo-Williams, M. T. Effectiveness of Student-Generated Video as a Teaching Tool for an Instrumental Technique in the Organic Chemistry Laboratory. *J. Chem. Educ.* **2016**, *93*, 141–145.
73. Hwang, C. S. Using continuous student feedback to course-correct during covid-19 for a nonmajors chemistry course. *J. Chem. Educ.* **2020**, *97*, 3400–3405.
74. Goff, L.; Knorr, K. Three heads are better than one: Students, faculty, and educational developers as co-developers of Science curriculum. *IJSaP* **2018**, *2*, 112–120.
75. Woolmer, C.; Sneddon, P.; Curry, G.; Hill, B.; Fehertavi, S.; Longbone, C.; Wallace, K. Student staff partnership to create an interdisciplinary science skills course in a research intensive university. *Int. J. Acad. Dev.* **2016**, *21*, 16–27.
76. Yilmaz, R.; Karaoglan Yilmaz, F. G. Assigned Roles as a Structuring Tool in Online Discussion Groups: Comparison of Transactional Distance and Knowledge Sharing Behaviors. *J. Educ. Comput. Res.* **2019**, *57*, 1303–1325.
77. De Wever, B.; Schellens, T.; Van Keer, H.; Valcke, M. Structuring Asynchronous Discussion Groups by Introducing Roles. *Small Group Res.* **2008**, *39*, 770–794.
78. Warren, A. N. Navigating assigned roles for asynchronous online discussions: Examining participants' orientation using conversation analysis. *Online Learn. J.* **2018**, *22*, 27–45.
79. Cook-Sather, A. Student-faculty partnership in explorations of pedagogical practice: a threshold concept in academic development. *Int. J. Acad. Dev.* **2014**, *19*, 186–198.
80. Dickerson, C.; Jarvis, J.; Stockwell, L. Staff–student collaboration: student learning from working together to enhance educational practice in higher education. *Teach. High. Educ.* **2016**, *21*, 249–265.

81. Barnes, E.; Goldring, L.; Bestwick, A.; Wood, J. A collaborative evaluation of student-staff partnership in inquiry-based educational development. *Staff-student partnerships in Higher Education* **2010**, 16–30.
82. Hattie, J. *Visible Learning: A Synthesis of Over 800 Meta-Analyses Relating to Achievement*; Routledge: 2008.
83. Creswell, J. W.; Poth, C. N. *Qualitative inquiry and research design: Choosing among five approaches*; Sage Publications: 2016.
84. Bovill, C.; Cook-Sather, A.; Felten, P. Students as co-creators of teaching approaches, course design, and curricula: implications for academic developers. *Int. J. Acad. Dev.* **2011**, *16*, 133–145.
85. Trowler, V. Student engagement literature review. *The Higher Education Academy* **2010**, *11*, 1–15.
86. Winstone, N.; Balloo, K.; Gravett, K.; Jacobs, D.; Keen, H. Who stands to benefit? Wellbeing, belonging and challenges to equity in engagement in extra-curricular activities at university. *Active Learn. High. Educ.* **2020**.
87. Mann, S. J. Alternative Perspectives on the Student Experience: Alienation and engagement. *Stud. High. Educ.* **2001**, *26*, 7–19.
88. Meyer, J.; Land, R.; Baile, C. *Threshold Concepts and Transformational Learning*; 2010.
89. Barnett, R.; Hallam, S. Teaching for supercomplexity: A pedagogy for higher education. In *Understanding pedagogy and its impact on learning*; Barnett R.; Hallam S., Eds.; 1999; p 137.
90. Bovill, C.; Cook-Sather, A.; Felten, P.; Millard, L.; Moore-Cherry, N. Addressing potential challenges in co-creating learning and teaching: Overcoming resistance, navigating institutional norms and ensuring inclusivity in student–staff partnerships. *High. Educ.* **2016**, *71*, 195–208.
91. Davis, B.; Sumara, D. Constructivist discourses and the field of education: Problems and possibilites. *Educ. Theory* **2002**, *52*, 409.
92. McInnerney, J. M.; Roberts, T. S. Online learning: Social interaction and the creation of a sense of community. *J. Educ. Technol. Soc.* **2004**, *7*, 73–81.
93. *Women, Minorities, and Persons with Disabilities in Science and Engineering: 2019*; NSF 19-304; National Science Foundation, Directorate for Social, Behavioral and Economic Sciences, National Center for Science and Engineering Statistics. https://ncses.nsf.gov/pubs/nsf19304/ (date accessed: October 2020).
94. Hess, C.; Gault, B.; Yi, Y. *Accelerating Change for Women Faculty of Color in STEM: Policy, Action, and Collaboration*; Institute for Women's Policy Research: 2013.
95. Joice, W.; Tetlow, A. ; *Disability STEM data for students and academic staff in higher education 2007/08 to 2018/19*; Report by Jisc for The Royal Society; 2021.
96. The Royal Society of Chemistry. *Diversity Data Report*; 2020.
97. The Royal Society of Chemistry. *Diversity landscape of the chemical sciences*; 2018.
98. *Metadata published by the UN Statistics Division collected from Annual Population Survey 2015 & Family Resources Survey 2015/16 (UK Data) and American Community Survey 2016 (US Data).* Data can be found at: https://unstats.un.org/unsd/demographic-social/sconcerns/disability/statistics/#/countries (date accessed: November 2020).
99. Cotton, D. R. E.; Joyner, M.; George, R.; Cotton, P. A. Understanding the gender and ethnicity attainment gap in UK higher education. *Innov. Educ. Teach. Int* **2016**, *53*, 475–486.

100. *Equality Challenge Unit and Advance HE, Equality in higher education: statistical report 2017*; 2017.
101. Crenshaw, K. Demarginalizing the intersection of race and sex: a black feminist critique of antidiscrimination doctrine, feminist theory and antiracist politics. *University of Chicago Legal Forum* **1989**, *1*, 8.
102. Liasidou, A. Intersectional understandings of disability and implications for a social justice reform agenda in education policy and practice. *Disabil. Soc.* **2013**, *28*, 299–312.
103. Equality Challenge Unit. *Experiences surrounding gender equality in the physical sciences, and their intersections with ethnicity and disability*; 2016.
104. *The UN Sustainable Goals Report 2020*. https://unstats.un.org/sdgs/report/2020/ (date accessed: May 2021).
105. Wilson-Kennedy, Z. S.; Payton-Stewart, F.; Winfield, L. L. Toward Intentional Diversity, Equity, and Respect in Chemistry Research and Practice. *J. Chem. Educ.* **2020**, *97*, 2041–2044.
106. Advance HE. *Tackling Racial Inequalities in Assessment in Higher Education A Multi-Disciplinary Case Study*, 2021.
107. Gravett, K. Troubling transitions and celebrating becomings: from pathway to rhizome. *Stud. High. Educ.* **2019**, 1–12.
108. Lygo-Baker, S.; Kinchin, I. M.; Winstone, N. E. *Engaging student voices in higher education: diverse perspectives and expectations in partnership*; Springer: 2019.
109. McLeod, J. Student voice and the politics of listening in higher education. *Crit. Stud. Educ.* **2011**, *52*, 179–189.
110. Balloo, K. In-depth profiles of the expectations of undergraduate students commencing university: a Q methodological analysis. *Stud. High. Educ.* **2018**, *43*, 2251–2262.
111. Lygo-Baker, S.; Kinchin, I. M.; Winstone, N. E. The Single Voice Fallacy. In *Engaging Student Voices in Higher Education : Diverse Perspectives and Expectations in Partnership*; Lygo-Baker, S., Kinchin, I. M., Winstone, N. E., Eds.; Springer International Publishing: Cham, 2019; pp 1–15.
112. Chen, W.; Wellman, B. The global digital divide–within and between countries. *IT & Society* **2004**, *1*, 39–45.
113. Eynon, R. Mapping the digital divide in Britain: implications for learning and education. *Learn. Media Technol.* **2009**, *34*, 277–290.
114. Gorski, P. Education Equity and the Digital Divide. *AACE Rev.* **2005**, *13*, 3–45.
115. Brown, N.; Nicholson, J.; Campbell, F. K.; Patel, M.; Knight, R.; Moore, S. Post-lockdown position paper. *Alter* **2020**, *15*, 262–269.
116. Meyer, A.; Rose, D. H.; Gordon, D. T. *Universal design for learning: Theory and practice*; CAST Professional Publishing: 2014.
117. Dinmore, S. P. The Case for Universal Design for Learning in Technology Enhanced Environments. *Int. J. Cyber Ethics Educ.* **2014**, *3*, 29–38.
118. Dewsbury, B. M. On faculty development of STEM inclusive teaching practices. *FEMS Microbiol. Lett.* **2017**, *364*.
119. Cohen, G. L.; Garcia, J. Identity, belonging, and achievement: A model, interventions, implications. *Curr. Dir. Psychol. Sci.* **2008**, *17*, 365–369.

120. Olga, T. Examining the relationship between student's engagement and socioeconomic background in higher education. *Student Engagement in Higher Education Journal* **2021**, *3*.
121. Cook-Sather, A. Increasing inclusivity through pedagogical partnerships between students and faculty. *Diversity and Democracy* **2019**, *22*.
122. Cook-Sather, A.; Krishna Prasad, S.; Marquis, E.; Ntem, A. Mobilizing a Culture Shift on Campus: Underrepresented Students as Educational Developers. *New Dir. Teach. Learn.* **2019**, *159*, 21–30.
123. Cook-Sather, A.; Agu, P. Student Consultant of Color and Faculty Members Working Together Toward Culturally Sustaining Pedagogy. *To Improve the Academy* **2013**, *32*, 271–285.
124. Cook-Sather, A.; Bovill, C.; Felten, P. *Engaging students as partners in learning and teaching: A guide for faculty*; John Wiley & Sons: 2014.
125. Delgado-Bernal, D. Theory, and critical raced-gendered epistemologies: Recognizing students of color as holders and creators of knowledge. *Qual. Inq.* **2002**, *8*, 105–126.
126. Heffernan, T. Sexism, racism, prejudice, and bias: a literature review and synthesis of research surrounding student evaluations of courses and teaching. *Assess. Eval. High. Educ.* **2021**, 1–11.
127. Perez, K. Striving Toward a Space for Equity and Inclusion in Physics Classrooms. *Teaching and Learning Together in Higher Education* **2016**.
128. White, K. N.; Vincent-Layton, K.; Villarreal, B. Equitable and Inclusive Practices Designed to Reduce Equity Gaps in Undergraduate Chemistry Courses. *J. Chem. Educ.* **2021**, *98*, 330–339.
129. Cohn, E.; Mullennix, J. W.; Branche, J. *Diversity across the curriculum: A guide for faculty in higher education*; Anker Pub. Co.: 2007.
130. Denson, N. Do curricular and cocurricular diversity activities influence racial bias? A meta-analysis. *Rev. Educ. Res* **2009**, *79*, 805–838.
131. Plaut, V. C. Diversity Science: Why and How Difference Makes a Difference. *Psychol. Inq.* **2010**, *21*, 77–99.
132. Good, J. J.; Bourne, K. A.; Drake, R. G. The impact of classroom diversity philosophies on the STEM performance of undergraduate students of color. *J. Exp. Soc. Psychol.* **2020**, *91*, 104026.
133. The 17 UN Sustainable Goals can be found at https://sdgs.un.org/goals, 2020 (date accessed December 2020).
134. Michalopoulou, E.; Shallcross, D. E.; Atkins, E.; Tierney, A.; Norman, N. C.; Preist, C.; O'Doherty, S.; Saunders, R.; Birkett, A.; Willmore, C.; Ninos, I. The End of Simple Problems: Repositioning Chemistry in Higher Education and Society Using a Systems Thinking Approach and the United Nations' Sustainable Development Goals as a Framework. *J. Chem. Educ.* **2019**, *96*, 2825–2835.
135. Goethe, E. V.; Colina, C. M. Taking Advantage of Diversity within the Classroom. *J. Chem. Educ.* **2018**, *95*, 189–192.
136. Hammond, Z. *Culturally responsive teaching and the brain: Promoting authentic engagement and rigor among culturally and linguistically diverse students*; Corwin Press: 2014.
137. Dunlop, L.; Hodgson, A.; Stubbs, J. E. Building capabilities in chemistry education: happiness and discomfort through philosophical dialogue in chemistry. *Chem. Educ. Res. Pract.* **2020**, *21*, 438–451.
138. Brown, D. R.; Brydges, S.; Lo, S. M.; Denton, M. E.; Borrego, M. J. A Collaborative Professional Development Program for Science Faculty and Graduate Students in Support of

Education Reform at Two-Year Hispanic-Serving Institutions. In *Best Practices in Chemistry Teacher Education*; American Chemical Society: 2019; Vol. 1335, pp 119–134.

139. Bovill, C. Co-creation in learning and teaching: the case for a whole-class approach in higher education. *High. Educ.* **2020**, 79, 1023–1037.

140. Woolmer, C. *Staff and students co-creating curricula in UK higher education: exploring process and evidencing value*; University of Glasgow: 2016.

141. Seale, J.; Gibson, S.; Haynes, J.; Potter, A. Power and resistance: Reflections on the rhetoric and reality of using participatory methods to promote student voice and engagement in higher education. *J. Furth. High. Educ.* **2015**, 39, 534–552.

Chapter 11

Options and Experiences for Online Chemistry Laboratory Instruction

Matt Morgan[*,1] and Emily Faulconer[1]

[1]Teachers College, Western Governors University, 4001 S. 700 E. #300, Salt Lake City, Utah 84107, United States
[2]Department of STEM Education, Embry-Riddle Aeronautical University, 1 Aerospace Boulevard, Daytona Beach, Florida 32614, United States
*Email: matthew.morgan@wgu.edu

While online course offerings have been on the rise for many years, there was a rapid transition to online courses due to campus closures and social distancing protocols associated with the COVID-19 pandemic. This has caused significant disruption in traditional chemistry laboratory activities. In this chapter, features of various distance laboratory approaches options are presented, including wet chemistry and dry laboratory options. Because the existing research comparing effectiveness of these strategies is scarce, no attempt is made to argue the effectiveness of one modality over the other, but rather advantages and disadvantages of each method are presented and discussed. The best fit for a specific institution will depend on many factors that must be considered.

Introduction

The laboratory experience is crucial to a student's chemistry education. Laboratory exercises serve several instructional functions, including reinforcement of concepts learned in lecture, the opportunity to apply knowledge, and development of laboratory and safety skills. Skills learned in the laboratory, such as solution preparation and data collection take chemistry beyond a purely academic exercise to a practical level of application.

In the last decade an increasing number of science courses have migrated to nontraditional formats, either fully online or using hybrid approaches, where the lecture portion takes place online and the laboratory experience occurs in person in a traditional laboratory. Fully online laboratory courses cover concepts and skills often applied in traditional in-person laboratory courses, though the nature of the experiment must be adjusted in order to perform experiments at the student's home, and potentially without supervision. Even traditional chemistry laboratory courses have adopted some non-traditional experiences such as remote equipment operation to varying degrees. Current global health concerns have precipitated the migration to these non-traditional laboratory learning environments. The COVID-19 pandemic and social distancing mandates have made instruction in the traditional laboratory setting difficult, if not impossible to accomplish. It is expected in a post-pandemic setting that online chemistry laboratory instruction will continue and the ideas discussed here will remain relevant.

Within higher education, some faculty and administrators suggest that laboratory courses cannot be offered remotely (1). Claims that chemistry cannot be taught online are often rooted in the claim of an inability to effectively teach the lab component outside of a traditional laboratory environment. However, the current literature supports the efficacy of the online laboratory instruction methods (2, 3). In the following sections, we will explore online and distance laboratory instruction options and review the literature on equivalence between traditional and nontraditional instruction techniques. Strengths and weaknesses of various laboratory modalities will be discussed, including issues like safety and student learning.

Laboratory Instruction Options

The literature does not present standard terminology for different teaching options for laboratory experiences. This chapter will define a traditional laboratory as one where students perform experiments in person in a physical laboratory. Wet chemistry activities include distance options where students make use of laboratory equipment, mixing chemicals and observing the results of those reactions. Dry lab options include simulations, virtual experiments, and remote laboratories, where the students do not handle chemicals themselves. Results for both these appraches can be reported online and the experiments can be performed synchronously or asynchronously.

There is no single answer for which laboratory approach works most effectively for every specific course or institution. The best instructional method depends upon specific factors, such as the primary purpose of the experiment, budget constraints, targeted rate of program growth and infrastructure capacity, and acceptable safety and access parameters. Faculty and administrators who are faced with making a chemistry laboratory modality decision may find a Lab Format Decision Matrix to be a helpful tool (4).

Similarly, the best-fit distance laboratory approach for students will depend upon student-specific external and internal factors. Sensory feedback may be more important to some students than others based on learning styles and preferences. Students' anxiety related to learning chemistry and handling chemicals is well known. It is unclear at this time how modality may influence student anxiety, as it is likely confounded by factors like math anxiety and computer anxiety. Currently, there is a gap in the literature to address this question, though it is likely that students will select online laboratory modalities through considering factors relevant to non-laboratory course modality decisions, including cost, with online courses typically being less expensive, time, and asynchronous flexibility for managing educational responsibilities with personal and professional commitments. Adequate workspace and technology requirements such as computer and internet availability may also be factors that influence student attitudes regarding laboratory instruction.

Distance Wet Chemistry Options

There are both positive and negative aspects to consider for distance curriculum that has students physically mixing chemicals when performing laboratory experiments. These features are summarized in Table 1 (5).

Table 1. A comparison of features of distance wet chemistry approaches

	Commercial Home Kits	Kitchen Chemistry
Tangible results with sensory feedback	✓	✓
Practical skill development	✓	✓
Low operating & maintenance costs for institution	✓	✓
Scalable (class size and sections)	✓	✓
Asynchronous 24/7 availability	✓	✓
Multiple access opportunities	✓	✓
Extended access time	✓	✓
Disability access	✓	✓
Student-instructor contact	(variable)	(variable)
Materials & Equipment Convenience	✓	
Additional student costs		✓
Replication of experiments		✓
Safety		

Both mail-order kits and kitchen chemistry provide lower operating and maintenance costs for the institution as there is no physical laboratory space to furnish and maintain. These methods also require no chemical inventory or institutional waste disposal. Both distance options are scalable for larger class sizes and more sections, where a traditional laboratory has limited time and physical space limitations. Students can benefit from the opportunity to engage with the material asynchronously, on their schedule rather than a fixed laboratory time. Multiple Access Opportunities refers to the ability of students to perform the exercises more than a single time, as there is often enough material to allow two or three experiment attempts. Students can also spend extended time working on an experiment or break the work into multiple sessions. This accessibility also extends to disability access to those with psychological or physical disabilities that limit or alter their ability to engage in a traditional laboratory.

A common argument in support of traditional laboratory experiences is the development of practical skills through hands-on work; distance laboratory options provide sensory feedback and allow for practical skill development in a way that online options cannot achieve. Because students are performing hands-on wet chemistry at home, they will need adequate space, ventilation, and safety equipment. Care and diligence must always be practiced by the student to use personal protective equipment like approved lab goggles and to prevent personal injury or damage to property. If the course is taught asynchronously, students will be working alone, without a lab partner, and without a lab instructor or teaching assistant to act as a safety observer. As with traditional laboratory courses, instructors teaching distance chemistry laboratory courses must attend to their duty of supervision and duty of care.

Instructor responsibilities that fall under duty of supervision include hazard communication, training in categories such as general emergency procedures and PPE use, waste management, and documentation. In a non-traditional laboratory, these duties could be upheld by infusing the course with laboratory safety skills and requiring students to demonstrate actionable safety knowledge in pre-lab assessments. Requiring students to submit a tour of their work area as an initial laboratory activity allows instructors to inspect housekeeping and chemical hygiene. Handling of wastes can be addressed both in pre-lab and post-lab activities. Instructors could even request video documentation of student procedures to provide feedback to refine laboratory skills and to ensure appropriate student behaviors. Including student compliance with health and safety policy as a graded course component could enhance documentation and provide a means of disciplinary action for noncompliance such as a reduction in assignment score or overall grade.

Instructor responsibilities that fall under duty of care include risk assessment and mitigation, generation of SOPs, consideration of PPE and engineering controls, emergency response, and documentation. Commercial companies and retailers may review their products for safety but they are held to different safety standards as they sell to the general public. To satisfy the duty of supervision, it is important that all experiments - whether materials are purchased as kits or not - are thoroughly reviewed for safety, including the generation of hazardous waste. When identifying laboratory experiments for distance students, perform rigorous hazards assessment If an institution has a safety review board, it may be beneficial to have the board review the selected laboratory experiments for use in the distance kits. Each laboratory experiment should have a standard operating procedure (SOP) developed and approved through the institution. According to OSHA's Laboratory Standard, the SOP should be reviewed annually. Many institutions offer template forms for hazard assessments and SOPs. Whether the SOP is administrative or is presented to students is an instructional design decision.

A passive approach by students will not work for distance lab activities. Students need to be engaged with the concepts and procedures and be able to troubleshoot problems that occur during the experiment. For this reason, students can benefit from options for reaching out to their instructor and their peers while performing an experiment at home. In synchronous courses, this can be achieved through various platforms, including Zoom, Skype, and Adobe Connect. In asynchronous courses, this can be achieved through platforms including the discussion forums within the learning management system and external options such as Google Hangout.

Commercial Home Kits

Students can purchase mail-order laboratory kits from a company like eScience Labs or MEL (6, 7). One logistical consideration with commercial kits is that instructors are reliant on companies to maintain appropriate inventories and shipping schedules. As many online and distance courses operate on a condensed semester schedule and shipping delays can quickly become problematic for students. A 2019 merger between home lab kit providers HOL and eScience combined with supply chain disruptions due to the COVID-19 pandemic continued to cause shipping delays (or shipment of incomplete kits) throughout 2020. Shipping to students living overseas can also be subject to delays due to international relations and global health measures. Equipment breakage can present additional complications for this approach. Glassware that arrives broken or breaks during the experiment are potentially hazardous and can be difficult to replace in time to allow students to meet assignment deadlines.

A key benefit of mail-order kits compared to kitchen chemistry is that they come with the chemicals and equipment necessary for students to perform experiments at home. Depending on the number of experiments in the course, the number of kit items can be overwhelming. Safety storage of chemicals and materials is another consideration. Students must take time to inventory the items and identify any pieces that are missing or broken. Having the instructor require this initial inventory and make it a graded exercise will increase the likelihood that the materials will be checked and deficiencies can be addressed quickly (8).

For safety and waste consideration, kits typically support microscale experimentation. For this reason, students often only have one opportunity to properly execute an experiment. If their work fails for any reason, they may be unable to generate the necessary data. This limitation is fairly unique to this lab option as traditional labs, kitchen chemistry, simulation, and virtual reality options offer easy avenues for replication of an experiment.

Many of the companies allow instructors to develop custom kits to align experiments with their curriculum. Instructors must choose from a limited menu of available experiments. Without a significant investment of time to develop in-house laboratory instructions, this can result in a piecemeal laboratory manual for students to follow. For example, a custom kit from e-Science may include experiments from two standard kits.

The laboratory manuals contain the full breadth of the experiments in the standard kits so either students have too much material to sift through or a trimmed version has inconsistent page numbers and experiment identification numbers. If instructors wish to have a more cohesive and polished presentation, they also must develop the laboratory manuals themselves. Creation of custom lab manuals requires a significant up-front investment of time, plus continued maintenance if the vendor changes lab equipment or procedures.

A post-lab discussion exercise could allow students to share experiences and collected data. This could enable students missing chemicals and equipment to still perform required calculations and draw conclusions from the exercise.

Kitchen Chemistry Experiments

The idea of at-home chemistry experimentation has been around for decades. If kitchen chemistry is used as a formalized approach within a nontraditional chemistry laboratory, a list of materials and lab procedures are provided to the student. The burden of collecting appropriate lab materials falls on the student, with purchases typically available through grocery and hardware stores. Online ordering may ease this burden, but availability will vary based on the geographic location of students. Unlike mail-order kits, purchases of chemical supplies are often in greater quantity than what is needed for experimentation. This provides some room for replication of experiments, though it generates an added risk for storage and disposal.

The internet abounds with ideas for kitchen chemistry experiments, from pen chromatography to the chemistry of rusting nails. Published literature on this approach is only starting to appear (*9*). Some websites offering other kitchen chemistry ideas include:

- The Royal Society of Chemistry's Kitchen Chemistry (*10*)
- Fizzics Education's Kitchen Chemistry Experiments (*11*)
- Science Center's Kitchen Chemistry Summer Camp Framework (*12*)
- American Chemical Society's Adventures in Chemistry (*13*)

When identifying potential kitchen chemistry experiments, instructors must consider appropriate rigor, affordable and accessible chemicals and materials, and safe chemical storage, use, and disposal. Just because a chemical is readily available does not mean that it is appropriate for at-home experimentation. For example, dissolving a nail with copper sulfate (available from the hardware store as a root killer for septic tanks) is not ideal as copper sulfate should be collected as hazardous waste due to its environmental toxicity (depending upon the concentration, volume, and institutional decisions regarding waste disposal).

Online Dry Lab Options

As with wet chemistry laboratory options, there are factors to consider when determining if a dry lab approach is workable for laboratory curriculum. Dry lab options, such as simulations, remote labs, and virtual reality, do not require students to manage physical inventory which greatly simplifies their laboratory experience. Many of these approaches require access to updated computer resources and the understanding of how to use them. There is also a learning curve for mastering the computers and the simulation software, and these hardware and software requirements can be barriers to learning (*14*). Another point to consider that dry lab experiments are not as immersive as their wet chemistry counterparts, but improvements and cost reductions in computer hardware and software should improve this aspect. As compared to learning via a distance wet chemistry approach, dry lab options provide the highest level of safety. These considerations are summarized in Table 2.

Table 2. A comparison of features of online laboratory modalities

	Laboratory Simulations	Remote Laboratory Experiences	Virtual Reality Experiments
Tangible results with sensory feedback			
Practical skill development		✓	(unknown)
Low operating & maintenance costs for institution	✓		✓
Scalable (class size and sections)	✓		✓
24/7 availability	✓	(variable)	✓
Multiple access opportunities	✓	✓	✓
Extended access time	✓	(variable)	✓
Disability access	✓	(variable)	✓

Table 2. (Continued). A comparison of features of online laboratory modalities

	Laboratory Simulations	Remote Laboratory Experiences	Virtual Reality Experiments
Student-instructor contact	(variable)	(variable)	(variable)
Materials & Equipment Convenience	✓		✓
Student costs	✓	✓	✓
Replication of experiments	✓	(variable)	✓
Safety	✓	✓	✓

Laboratory Simulations

Simulated chemistry experiments can be used for both traditional and nontraditional laboratory experiences (15). Simulations mimic the laboratory environment and have students manipulate digital versions of laboratory equipment and materials. Distinct from laboratory simulations, there are many interactive elements available that aim to teach specific chemistry concepts at the macro and micro level, including PhET interactive simulations, MERLOT Chemistry Simulations, and ChemReaX (16–18). Some sources, like ChemCollective, offer virtual lab experiments, simulations, and molecular level visualizations (19). Commercial options for laboratory experiment simulations include Labster and LateNiteLabs (20, 21).

While virtual labs offer safety, they often ignore safety skills. For example, within LateNiteLabs, students cannot accidentally spill chemicals and when they are done with an experiment, they discard chemical labware and chemicals into a trash bin, which in no way simulates an actual laboratory environment. Students who progressed to a hands-on second-semester laboratory course would be under equipped to deal with chemical and physical hazards present when working with chemicals. Simulations also lack realism in other ways. They often do not provide variability of results and present occasional random outcomes. One benefit of simulations, though, is also safety. Simulations are ideal for allowing students to experience laboratory activities that are simply too dangerous for the average undergraduate student to execute in a wet chemistry experiment, and this is true for both in a traditional or nontraditional lab settings.

Remote Laboratory Experiences

Data acquisition through remote equipment operation is an option that has been used for decades in traditional laboratories, though often for advanced laboratory operations like atomic absorption spectrometry (22). Furthermore, there is complexity in scheduling student remote access and shipping samples to the instrumentation for data acquisition. The dominating factor for remote laboratories, however, is cost. The Colorado Community College System received external funding to set up and run a remote laboratory for students studying chemistry, physics, and biology. Remote experiments were effective for student learning, but the arrangement was not financially sustainable. The remote laboratory closed when the external funding ran out (23).

Virtual Reality Experiments

Virtual reality (VR) laboratory experimentation is an emerging modality option. Common applications of VR include exploration of chemistry concepts on the microscopic/submicroscopic scale (molecular shape, bonding, etc.), but there is an increasing attention to the use of this technology for macroscopic exploration through experiments and experiences. Students can move around a three-dimensional virtual laboratory space through digitized video and animations of experiments, creating an interactive and immersive experience. VR experiences have varying levels of reality, ranging from idealized or abstracted experiences to fully interactive visualizations (24). The use of fixed narrative structures or fixed viewing perspectives limits interactivity and active learning (25). The advancements in affordable high-quality 360° video recording technology has made fully interactive visualizations easier to achieve.

VR laboratories can be deployed with 1) readily-available portable headset hardware like Google cardboard device and app or 2) high-performance PC-tethered technology. The portable option would have low cost to students to purchase the cardboard kit but would require access to a high-quality smartphone. The high-performance option would require hardware not likely to be possessed by students in a distributed (off-campus) model and would require coordination with an institution's information technology team to establish for on-campus use (26).

Because VR experiences are so novel in higher education, it may be challenging to seamlessly integrate into the curriculum with appropriate scaffolding for students. Disability access must also be considered, including alternative options for manipulating the interactive environment. Students will have a range of spatial skill proficiency; alternative options can ensure equitable learning opportunities. Ergonomics is another potential limitation of VR hardware such as use for students who wear glasses.

Comparison of Laboratory Choices

Currently, the research comparing chemistry laboratory modalities is very limited. With the rapid transition to online in the wake of COVID-19, it is likely that there will be more research in this area moving forward. Emerging data suggests that chemistry laboratory courses are amenable to online learning in higher education and can promote achievement of student learning outcomes at least equivalent to traditional laboratory environments.

Comparisons for single-experiments offered as an online simulation versus in-person showed that simulations were equivalent to or better than in-person experiences for student performance and content knowledge mastery (27). No difference in practical motor skill development has been reported and one study reported higher student performance in a simulated lab compared to a teacher-centered, expository lab (28).

Comparison for an entire course offered using laboratory kits versus in-person showed no difference in pass rate or withdrawal rate but did report a positive skew in grade distribution for students using the lab kits compared to in-person experimentation (29). At the secondary level, a lab course using simulations has been shown to be as effective as a traditional in-person laboratory experience (30). When comparing single experiments offered as VR versus in-person, no difference in long term content knowledge retention was shown, but students did self-report stronger learning with VR relative to 2D text and images (31).

There are many aspects in comparing laboratory modalities that are as-yet unexplored. While VR, lab kits, and in-person laboratories require motor skills and application of practical laboratory skills, it is unclear at this time if skill development is equivalent across modalities. It is unclear if student actionable safety knowledge is equivalent across laboratory modalities. It is also unclear how modality may influence chemistry and laboratory anxiety.

Conclusions

At this time, there is not enough evidence to definitively evaluate equivalence between these different laboratory options in areas including mastery of disciplinary knowledge, laboratory skills, actionable safety knowledge. Student opinions and attitudes towards learning chemistry by different online methods also remains under-explored. However, there are certain features of these strategies that make one approach more viable than another for a given institution. Understanding the benefits and disadvantages can help administrators, curriculum developers, and educators evaluate which online modality may work best in their setting and for their students.

References

1. Vutukuru, M. Faulty assumptions about lab teaching during COVID. *Inside Higher Ed.* https://www.insidehighered.com/advice/2020/08/05/engineering-instructor-disagrees-notion-lab-courses-cant-be-taught-effectively (accessed Oct. 5, 2020).
2. Casanova, R. S.; Civelli, J. L.; Kimbrough, D. R.; Heath, B. P.; Reeves, J. H. Distance learning: a viable alternative to the conventional lecture-lab format in general chemistry. *J. Chem. Educ.* **2006**, *83*, 501; DOI: 10.1021/ed083p501.
3. Mawn, M. V.; Carrico, C.; Charuk, K.; Stote, K. S.; Laurence, B. Hands-on and online: scientific explorations through distance learning. *Open Learning: The Journal of Open, Distance and e-Learning* **2011**, *26*, 135–146; DOI: 10.1080/02680513.2011.567464.
4. Faulconer, E. K.; Faulconer, L. S.; Hanamean, J. R. *Arriving at a better answer: a decision matrix for science lab course format*; National Science Teaching Association. https://www.nsta.org/arriving-better-answer-decision-matrix-science-lab-course-format (accessed Nov. 15, 2020).
5. Faulconer, E. K.; Gruss, A. B. A review to weigh the pros and cons of online, remote, and distance science laboratory experiences. *IRRODL* **2018**, *19*; DOI: 10.19173/

irrodl.v19i2.3386. https://books.google.com/books?hl=en&lr=&id=T0-a1Q-0chgC&oi=fnd&pg=PA83&dq=kitchen+chemistry&ots=8Xj62v9m6q&sig=_WIbYIKU5iYOpQQqQnPS4TWGaiE. (accessed Apr. 24, 2021)

6. *eScience Labs.* https://esciencelabs.com/ (accessed Dec. 15, 2020).
7. *MEL Science.* https://melscience.com/US-en/ (accessed Dec. 15, 2020).
8. Reeves, J.; Kimbrough, D. Solving the laboratory dilemma in distance learning general chemistry. *Journal of Asynchronous Learning Networks.* **2004**, *8*, 47–51. https://pdfs.semanticscholar.org/72e6/0c959d02b57faaa791ff0e131901180aa948.pdf (accessed Apr. 24, 2021)
9. Andrews, J. L.; de Los Rios, J. P.; Rayaluru, M.; Lee, S.; Mai, L.; Schusser, A.; Mak, C. H. Experimenting with At-Home General Chemistry Laboratories During the COVID-19 Pandemic. *J. Chem. Educ.* **2020**, *97*, 1887–1894; DOI: 10.1021/acs.jchemed.0c00483.
10. Royal Society of Chemistry. *Kitchen chemistry.* https://edu.rsc.org/resources/collections/kitchen-chemistry (accessed Dec. 15, 2020).
11. *Fizzics Education.* https://www.fizzicseducation.com.au/category/150-science-experiments/kitchen-chemistry-experiments/ (accessed Dec. 15, 2020).
12. Science Center, *Kitchen chemistry.* http://www.sciencenter.org/chemistry/d/framework_kitchen_chemistry.pdf (accessed Dec 15, 2020).
13. *ACS Adventures in Chemistry.* https://www.acs.org/content/acs/en/education/whatischemistry/adventures-in-chemistry.html (accessed Dec. 15, 2020).
14. Ali, N.; Ullah, S. Review to analyze and compare virtual chemistry laboratories for their use in education. *J. Chem. Educ.* **2020**, *97*, 3563–3574; DOI: 10.1021/acs.jchemed.0c00185.
15. Yaron, D.; Karabinos, M.; Lange, D.; Greeno, J. G.; Leinhardt, G. The ChemCollective—virtual labs for introductory chemistry courses. *Science* **2010**, *238*, 584–585; DOI: 10.1126/science.1182435.
16. *PhET Interactive Simulations.* https://phet.colorado.edu/ (accessed Feb. 3, 2021).
17. *MERLOT Virtual Labs.* https://virtuallabs.merlot.org/vl_chemistry.html#MERLOT (accessed Feb. 3, 2021).
18. *ChemReaX.* http://www.sciencebysimulation.com/chemreax/Analyzer.aspx (accessed Feb. 3, 2021).
19. *ChemCollective.* http://www.chemcollective.org/ (accessed Feb. 3, 2021).
20. *Labster.* http://labster.com (accessed Jan. 25, 2021).
21. *LateNiteLabs.* https://www.studica.com/it/en/late-nite-labs-virtual-science (accessed Jan. 25, 2021).
22. Erasmus, D. J.; Brewer, S. E.; Cinel, B. Assessing the engagement, learning, and overall experience of students operating an atomic absorption spectrophotometer with remote access technology. *Biochemistry & Molecular Biology Education* **2015**, *43*, 6–12; DOI: 10.1002/bmb.20838.
23. Bates, T. Can you teach lab science via remote labs? *Online Learning and Distance Education Resources*, 2013. https://www.tonybates.ca/2013/04/22/can-you-teach-lab-science-via-remote-labs/ (accessed Feb. 5, 2021).
24. *Oxford VR Group.* http://www.chem.ox.ac.uk/vrchemistry (accessed Dec. 15, 2020).

25. Limniou, M.; Roberts, D.; Papadopoulos, N. Full immersive virtual environment CAVE™ in chemistry education. *Computers and Education* **2008**, *51*, 584–593; DOI: 10.1016/j.compedu.2007.06.014.
26. Dunnagan, C. L.; Dannenberg, D. A.; Cuales, M. P.; Earnest, A. D.; Gurnsey, R. M.; Gallardo-Williams, M. T. Production and evaluation of a realistic immersive virtual reality organic chemistry laboratory experience: infrared spectroscopy. *J. Chem. Educ.* **2020**, *97*, 258–262; DOI: 10.1021/acs.jchemed.9b00705.
27. Hawkins, I.; Phelps, A. J. Virtual laboratory vs. traditional laboratory: which is more effective for teaching electrochemistry? *Chem. Educ. Res. Pract.* **2013**, *14*, 516–523; DOI: 10.1039/C3RP00070B.
28. Pyatt, K.; Sims, R. Learner performance and attitudes in traditional versus simulated laboratory experiences. In ICT: Providing choices for learners and learning. *Proceedings ascilite Singapore 2007*. https://citeseerx.ist.psu.edu/viewdoc/download?doi=10.1.1.513.3544&rep=rep1&type=pdf (accessed Apr. 24, 2021).
29. Faulconer, E. K.; Griffith, J. C.; Wood, B. L.; Acharyya, S.; Roberts, D. L. A comparison of online and traditional chemistry lecture and lab. *Chem. Educ. Res. Pract.* **2018**, *19*; DOI: 10.1039/c7rp00173h.
30. Tatli, Z.; Ayas, A. Effect of a Virtual Chemistry Laboratory on Students' Achievement. *Journal of Educational Technology & Society* **2013**, *16*, 159–170. http://ezproxy.libproxy.db.erau.edu/login?url=https://www-proquest-com.ezproxy.libproxy.db.erau.edu/scholarly-journals/effect-virtual-chemistry-laboratory-on-students/docview/1287029545/se-2?accountid=27203.
31. Qin, T.; Cook, M.; Courtney, M. Exploring chemistry with wireless, PC-less portable virtual reality laboratories. *J. Chem. Educ.* **2021**, *98*, 521–529; DOI:10.1021/acs.jchemed.0c00954.

Chapter 12

Lessons Learned from Implementing Blended and Online Undergraduate Chemistry Laboratory Teaching during the Covid-19 Pandemic

Helen Cramman,[1,*] Mia A. B. Connor,[2] Chapman Hau,[2] and Jacquie Robson[2]

[1]School of Education, Durham University, Leazes Road, Durham DH1 1TA, United Kingdom
[2]Department of Chemistry, Durham University, Science Site, Stockton Road, Durham DH1 3LE, United Kingdom
*Email: helen.cramman@durham.ac.uk

In March 2020, the Covid-19 pandemic led to unprecedented circumstances which impacted significantly on Higher Education. Since that time, requirements for social distancing and reduced access to in-lab teaching facilities have meant a dramatic redesign of many Chemistry undergraduate laboratory courses. This chapter presents the lessons learned from the redevelopment of the 2020-2021 first-year chemistry undergraduate laboratory course at Durham University. The two pre-existing laboratory modules were converted from their traditional in-lab delivery (supported by online pre- and post-lab activities) to a blended delivery module and a fully online module. The blended module focused on the key manipulative skills students need to gain competence in to progress successfully to second year laboratory work. The fully online module focused on scientific enquiry skills. This chapter presents practical and theoretical considerations for the development of blended or online laboratory courses before discussing lessons learned from the evaluation of the process of implementing the course and the impact for students.

Introduction

In March 2020, the Covid-19 pandemic forced a change to the way that undergraduate laboratory courses were taught. Although this was not a strategically planned change, it nevertheless presented an opportunity for rethinking and redeveloping laboratory teaching using different modes of delivery (Table 1), which otherwise may have not been possible.

© 2021 American Chemical Society

Table 1. Definitions for the modes of laboratory delivery

Mode of delivery	Definition for this chapter
Distance/remote Learning (DL)	An approach to teaching and learning in which there is a physical, and often psychological, separation between students and instructors (1).
Online learning (OL)	Learning experiences in synchronous or asynchronous environments using different devices (e.g., mobile phones, laptops, etc.) with internet access (2).
Blended learning (BL)	The combination of in-person and online learning activities (3).

However, the situation presented by Covid-19 also produced several challenges: 1) that students had not applied for their degree courses expecting to undertake laboratory work remotely 2) that the time for planning and redeveloping materials by course instructors was limited and 3) that the learning objectives for the course could not be changed as these are agreed significantly in advance of the timeframe which was available for redevelopment.

The approach taken for the first year Chemistry undergraduate laboratory course at Durham University was to rethink which key aspects of the course required in-person "hands-on" laboratory time and which could be effectively taught online. The changes were subsequently evaluated to understand the impact for students and staff.

In this chapter we present an overview of evidence in the literature to support the decision-making for the course redevelopment, followed by a detailed description of the design of the new course, evidence from the evaluation, concluding with lessons learned and challenges for the future of laboratory teaching.

Considerations for the Design of Laboratory Courses

The following section summarises factors for consideration in course design, split into three areas:

1) Identifying learning objectives
2) Creating an environment for meaningful learning
3) Understanding student progress

1) Identifying Learning Objectives

The focus of in-person laboratory work has traditionally been on the development of psychomotor skills (i.e. hands-on manipulation of equipment, instruments and chemicals) (4). In the UK, the Quality Assurance Agency (QAA) and Royal Society of Chemistry (RSC) have outlined benchmark standards which detail the levels of competence expected for holders of chemistry degrees (5). Pre-Covid-19, the RSC accreditation guidance required that "Students must develop a range of practical skills" and included reference to a typical time requirement of 400 hours of in-lab time for an integrated Masters degree, with 300 hours for a BSc (6), with little specific detail about the exact types of skills which must be covered by the course. The guidance was then interpreted by course designers to identify the key areas and competences that were to be addressed within modules. Table 2 shows the subdivision of chemistry practical work competences into three skill groups to support implementation within laboratory courses.

Table 2. Subdivision of chemistry practical work competences

Skill type	Definition	Further division
Hard	Skills related to the technical aspects to undertake tasks, often taking into account the acquisition of knowledge (8). Usually highly specific, hands-on skills (e.g. using apparatus).	Psychomotor skills (manipulative) Procedural skills (having the knowledge to successfully use certain techniques and implement them) General laboratory skills (e.g. knowing how to act safely in a laboratory)
Soft	Relate to the "behaviours that promote the formation of skills applied to acquire knowledge and then disseminate what is obtained" (9). Subject knowledge is required for these skills but they are often more subjective than hard skills in the way they are assessed and are usually developed rather than taught.	Data-use skills Enquiry-based skills Communicating science skills.
Transferrable	Skills which go beyond the laboratory and can be utilised in a range of different fields. Often defined as "the interpersonal, human, people, or behavioural skills needed to apply technical skills and knowledge (10)" (e.g. organisational, communication and self-motivation skills) (11).	

Due to the unprecedented situation created by Covid-19, the RSC issued updated guidance in September 2020 related to the requirements for laboratory work during the pandemic, stating that online resources could be used to ensure that students are still exposed to a wide range of practical techniques, where in-lab work is not possible (7). The existing guidance on number of hours of in-lab time was also no longer considered a feasible requirement and flexibility would be used by accreditation panels, with a focus on the skills students obtained and not solely the lab hours. The specific skills were not described.

2) Creating an Environment for Meaningful Learning

For meaningful learning to occur in the undergraduate chemistry laboratory, psychomotor experiences should be assimilated with the cognitive and affective domains, i.e. manipulative skills should be combined with the "thinking behind" and "feeling of" doing science (12). An example of a task that could promote the "thinking behind the doing" could be asking students to deduce and explain why reagents must be added in a specific order to prevent the formation of unwanted byproducts during a synthesis. This task accompanies the in-lab activity, where the synthesis is carried out.

Opportunities should be provided for students to relate new knowledge to relevant concepts and propositions (13).

Consideration of several factors have been shown to facilitate a meaningful learning environment.

- *Interactions* (Figure 1) - Key to learning being achieved through interactions is the bidirectional flow of information. Online social interactions may be synchronous, when communication takes place instantaneously (e.g. live chat and videoconferencing) or asynchronous via delayed communication (e.g. discussion boards and email) (*14*).
- *Attitude* - Attitude is comprised of three components: behavioural, cognitive and affective components (*15*). Recent research has indicated that students consider online learning to be the least effective and least enjoyable method of learning, and can lead to lower affective student attitudes (*16*).
- *Self-efficacy* – This is a student's judgement of their own capability (*17*), developed by four main sources of influence: mastery experiences, vicarious experiences, social persuasion and emotional states (*18, 19*).
- *Confidence* – A student's feeling of assuredness and lack of anxiety when completing laboratory activities (*20*). Pre-laboratory activities (including the use of videos, quizzes, simulations and exercises) have been reported to increase students' confidence in completing practical activities (*21–24*).

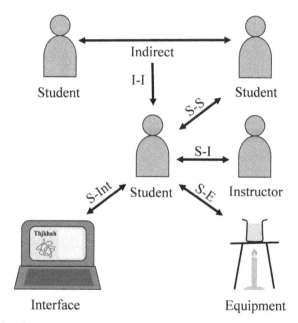

Figure 1. Summary of student interactions in a chemistry laboratory including students engaging with other people (student-student, S-S or student-instructor, S-I), with equipment (student–equipment, S-E) and indirect interactions (I-I) e.g. overhearing conversations of others (25). Student-interface interactions (S-Int) are distinctive to distance learning (26) and use technology to mediate interactions between students, instructors and content.

3) Understanding Student Progress

Effective laboratory work requires links between:

1. the laboratory task (i.e. what students are intended to do) and laboratory actions (i.e. what students actually do)

2. the laboratory objectives (i.e. what students are intended to learn) and the laboratory learning (i.e. what students actually learn).

These can be evidenced through whether a student both finishes the task *and* whether they fulfil the stated key learning outcomes (27).

The Redeveloped First-Year Undergraduate Chemistry Laboratory Course at Durham University

Pre-Covid, the Practical Chemistry 1A (P1A) and Practical Chemistry 1B (P1B) first year modules at Durham University were run as in-lab practical chemistry courses with online support (via the Virtual Learning Environment, VLE) both before and after each laboratory session. In some institutions the VLE may be referred to as the Learning Management System (LME). VLE will be used to refer to both in this chapter. The modules ran over two 10-week terms.

Due to social distancing requirements in the 2020/21 academic year reducing the number of students able to be in the lab at the same time from 60 to 18, the course was redesigned so that the two modules focused on different skills and were delivered via different methods. The aim of the redeveloped lab course was to achieve the same learning outcomes as the pre-Covid course, using implementation methods which were compliant with Covid-19 restrictions.

P1A was redeveloped as a blended laboratory module with alternating complementary in-lab sessions and online activities. The redesigned module focused on covering the key manipulative 'hard' skills students would need to gain competence to progress successfully to second year laboratory work, with the online component complementing this by delivering the related key procedural 'hard' and 'soft' skills (Figure 2). Students were scheduled to attend an in-lab session every fortnight (reduced one session per week prior to Covid-19) and online exercises and activities in the week in-between. Due to further Covid-19 restrictions, students were only able to undertake a maximum of three out of the planned nine P1A in-lab sessions and the remainder moved to online activities.

P1B was redeveloped to be delivered entirely online and focused on procedural 'hard' skills, 'soft' skills (including scientific enquiry skills using data sets for analysis, theoretical situations and experimental planning activities), and 'transferable' skills.

By the end of the course in March 2021, P1A students were expected to have completed three in-lab and 12 online activities and P1B students to have completed 16 online activities.

The Aim and Design of the Redeveloped Lab Course

Given the lack of access to the lab, the development of hands-on, manipulative 'hard' skills was restricted. The goal of the online activities was to enable students to focus on the 'thinking behind the doing' (i.e. how, when and why different skills are used) until they could be in the lab to practice them along with analysing data and completing the usual 'out-of-lab' aspects of the course. The aim was to reduce the cognitive load on students when they were able to undertake work in the laboratory. The limited in-lab sessions aimed to allow students to experience as many of the manipulative skills covered in the course as possible. Figure 2 shows the structure and timeline of the P1A and P1B modules.

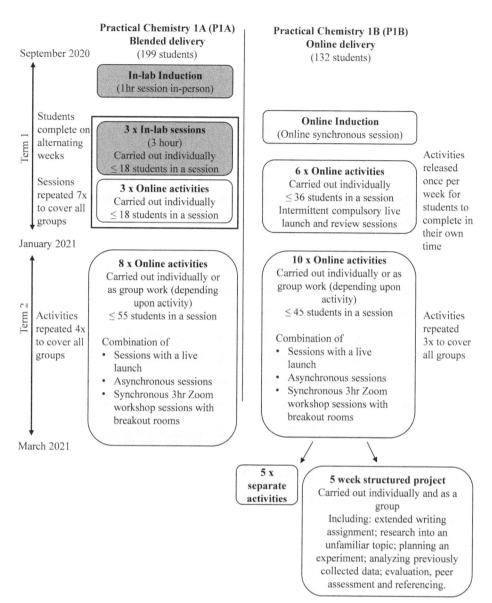

Figure 2. Structure of the P1A and P1B modules, including in-lab sessions in dark grey and online activities in light grey.

When working online, P1A students were encouraged to undertake their work during their timetabled session for the module, however, they could choose to complete the work flexibly at a time which best suited them during the two-week timeframe for the activity. Zoom breakout room workshops were introduced to both P1A and P1B in term 2 to facilitate interaction between students, as well as synchronous interaction with teaching staff. The workshop sessions were designed to provide students with the opportunity to form social work groups as well as academic support. Breakout room groups were randomly allocated so students met new peers each time. Figure 3 shows the materials, resources and support arrangements for the in-lab and online activities.

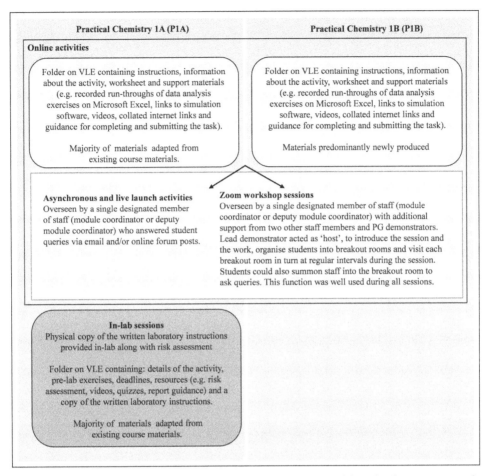

Figure 3. Materials, resources and support arrangements for the P1A in-lab and online activities and P1B online activities.

Where possible, feedback was collected over the duration of the course from students and staff (including the course coordinator, demonstrators and advisors) and was used to update and adapt course materials and delivery. Adaptions included the introduction of the Zoom workshops to tackle student complaints of feeling isolated or not knowing anyone on their course. Adaptions also had to be made to ensure compliance with changing restrictions due to Covid-19.

Method

Data were collected across three research studies (Table 3) to understand both the context of the incoming first year chemistry undergraduate cohort in October 2020 and to evaluate the skills development and experiences of students in their laboratory modules at the end of term 1. One Zoom workshop had been held at the point of data collection for study 2 and 3.

Table 3. Research studies with the 2020/21 first year undergraduate chemistry cohort

Study	Method	Cohort	No. of participants taking part in the study
Study 1	Survey 1 (Sept 2020)	210 students	103 (49%)
Study 2	Survey 2 (Jan/Feb 2021)	199 students (45.2% female, 54.8% male)	122 students (61% response rate) (49.5% female, 48.5% male, 1.8% prefer not to say)
	Semi-structured interviews (Feb 2021)	Volunteers through the survey	9 students (5 male, 4 female)
	Semi-structured interview (Feb 2021)	Course coordinator	1 member of staff
Study 3	Survey 3 – students and demonstrators (Jan/Feb 2021)	199 students (45.2% female, 54.8% male)	86 students (43% response rate) (52% female, 44% male, 3% prefer not to say)
		16 demonstrators	8 demonstrators (50% response rate) (63% female, 25% male, 13% prefer not to say)

Results

Successful Completion of Work and Meeting Learning Outcomes

Of the 86 (43%) students responding to survey 3 relating to the P1A module, 45% strongly agreed that they had successfully completed their in-lab work, with 36% somewhat agreeing. The figure was less for P1A online activities, with 21% strongly agreeing and 55% somewhat agreeing that they had successfully completed their work. Module marks were not available at the time of data collection due to covid-related assessment timeline changes.

Interactions

Students were asked to report on the frequency with which they participated in a specified set of interactions in the in-lab and online activities (Table 4). A distinct difference was seen between communication with the demonstrator between the in-lab sessions and online activities (survey 3). When asked about materials for supporting their learning (survey 2), 34% of students found the videos for both the in-lab and online activities very useful.

Table 4. Interactions which the greatest percentage of students carried out more than three times per in-lab or online session.

Interaction	% respondents reporting carrying out interaction more than 3 times per session
In-lab sessions	
Reading the written instructions	64%
Talking to a demonstrator about lab equipment	31%
Talking to a demonstrator about lab procedures	31%
Online activities	
Observing or listening to a video from a demonstrator providing guidance on the lab task	45%
Observing or listening to a video demonstration of how to do a lab technique	43%
Using the internet to look up information about the lab task	43%
Online activity lowest %: Communicating with a demonstrator about the task instructions, general guidance on the task, scientific concepts, equipment or techniques or data analysis	≤2%

The students predominantly worked individually for the in-lab and online activities in term 1. Twenty-three percent of respondents (survey 3) reported they had not completed any online activities with another student or group of students (data were collected prior to students attending the Zoom workshop sessions).

All the demonstrators that responded to survey 3 reported that they were more likely to initiate interactions with students first in an in-lab context than the students initiating the contact. However, the result was split 50:50 in the online context, with half saying that they would wait for students to initiate the interaction. The four respondents who said that they would wait for the student to initiate the interaction first in the online context had all been demonstrating for more than three years.

Attitudes

Respondents to survey 2 rated on a scale of 1 to 5 how they felt about in-lab and online activities (Figure 4). A significant difference was observed between students' perceptions as to how worthwhile in-lab and online activities were. Eighty percent 80% of respondents gave a score of 5 (worthwhile) for in-lab sessions compared to only 17% for online activities. No respondents rated online activities as being at the top of the scale for excitement, compared to 37% for in-lab.

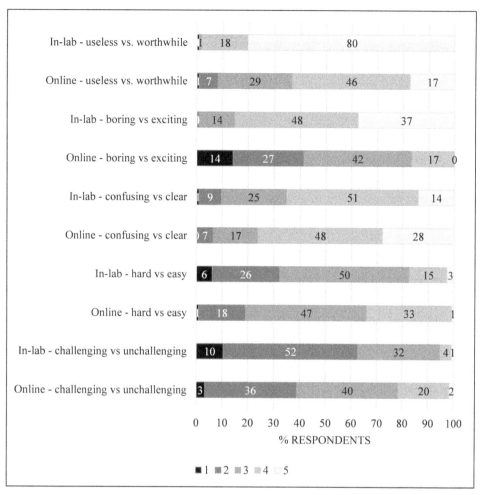

Figure 4. Percentage of respondents choosing ratings from 1 to 5 for their perceptions about in-lab and online activities (1 = leftmost description to 5 = rightmost description e.g. 1 – useless, 5 – worthwhile, n = 115 – 122)

Respondents in study 2 expressed that they felt they were missing out on developing 'practical' skills due to undertaking fewer in-lab sessions than they had been expecting when they applied for their degree course. Participants also indicated that they were worried that they would be unprepared for future years of study due to the loss of in-lab time. The respondents did not consider that online activities replaced their expectations of the skills they would have developed within an in-lab session.

"I just don't think that a practical course can be replaced with something that's not practical... Actually putting acid in the flask and watching something happening and doing something, it's just you can't compare."

Respondents enjoyed in-lab activities as they considered that it gave them the opportunity to gain hands-on practical experience. This met with their expectations for a chemistry degree.

At the start of the second term, online simulations were introduced to the course. Students' attitudes were mixed towards the introduction of the simulations with some enjoying these and others not seeing how they helped to develop lab skills.

Self-Efficacy and Confidence

Respondents to survey 2 reported that the online activities typically took longer to complete (Table 5), however, it should be noted that pre- and post-lab activities were not included in the time for in-lab sessions. Only 8% of respondents completed all the online activities in the time-tabled slot, with 28% reporting that they had never completed them in this time slot.

Table 5. Time taken to complete in-lab and online sessions. (n = 112 in-lab, n = 114 online)

Mode	Time to complete session	% respondents
In-lab	1 – 2 hours	15
	2 – 3 hours	85
Online	1 – 2 hours	26
	2 – 3 hours	54
	3 – 5 hours	19

Feedback from the course coordinator indicated that in order for the students to effectively engage with the online lab course, the students had to understand how to use the VLE, had to ensure that they checked the weekly schedule at the start of each week and read emails and announcements. These time management and organisation skills were essential for effective engagement with the course.

Table 6 shows the skills (from a specified list of 32 skills) which more than 60% of respondents to survey 2 indicated that they could carry out well and which they were confident in undertaking after the first term of teaching (indicated by selecting the highest rating of 5). A strong correlation was observed ($r = 0.973$, $p<0.001$) between how well students considered they could carry out the 32 skills and how confident they were.

Table 6. The skills which more than 60% of respondents to survey 2 gave the highest score of 5 for ability or confidence. The survey contained a list of 32 specified skills for students to provide a rating of 1 (lowest) to 5 (highest) for their self-assessed ability and confidence. (n = 101 – 104)

Skill	Type of skill	How well (%)	How confident (%)
Using a top-pan balance	Manipulative	83	80
Washing glassware	Manipulative	69	67
Using Microsoft Excel to calculate the mean of a dataset	Soft	68	62
Using appropriately sized measuring cylinders to measure volumes	Procedural	65	55
Using cell references in Microsoft Excel	Soft	63	62
Using Microsoft Excel to calculate the standard deviation of a dataset	Soft	60	50
Using formula in Microsoft Excel	Soft	56	64

Zoom Workshops

At the time of the interviews in study 2, when participants had completed one online Zoom workshop, they considered the group work that had been introduced through this method had overcome some of the challenges faced by completing online activities independently. The students particularly liked the opportunity to get to know others on their course, and that they were able to discuss thoughts and ideas with other students. It also reduced their sense of being alone.

However, some students commented that not all students participated equally in the workshop sessions and that conversation could be dominated by a small number of individuals. They suggested that the lack of social cues relating to when other students were going to speak and taking time to become comfortable sharing ideas may have impacted. Having the same group of students in a breakout room group in each session, rather than changing groups each time, was suggested as a way to overcome this. Feedback from staff indicated that some students were extremely anxious about attending the Zoom workshops. Accommodations were therefore made so that students could attend only the introduction to the session with camera and microphone off and then leave to complete the session individually, rather than in a group with other students.

The course coordinator commented that when they "dropped into" the breakout rooms in the Zoom workshops, the students asked more questions than they typically did in pre-Covid in-lab sessions. They postulated that this may have been due to the online interface feeling less intimidating than asking questions in person in the lab.

What Worked Well?

The flexibility of being able to work at their own pace, in their own time, so long as they submitted within the deadlines, was considered to be a particular benefit of the online activities by students. The clarity of instructions was also commented as having been particularly good. Respondents considered that they had been taught a range of different skills in the online laboratory activities and that these skills that were not always developed as well in the in-lab environment.

Challenges and Barriers

Technological Difficulties

Twenty-five percent of respondents to survey 2 reported experiencing technical difficulties, with issues with internet connection affecting their ability to complete online activities. The course coordinator reported that managing the Zoom workshops was made more difficult by students having connectivity issues. This was disruptive to discussions between the staff 'host' and the other student groups in the breakout rooms as the member of staff had to leave to re-admit the students with the lost connection to the Zoom session and then re-allocate them to their breakout rooms. There had also been connectivity issues for the course coordinator which disrupted one synchronous session.

Impacts of the Covid-19 Pandemic

Participants in study 2 reported feeling 'isolated' when completing both in-lab and online activities due to social distancing requirements, including the requirement to wear face masks in the in-lab sessions. The requirement for social distancing since the beginning of their time at university meant that students felt that they knew few peers on their course and felt like it was harder to interact with other students both in-lab and online. This also impacted on how confident they felt being able to ask for help. The course coordinator had noted that they had not received much informal feedback from students during the activities and that the online environment had made gauging how students were feeling, by observing their body language and via informal chat, very difficult. They reflected that they had not realised until this year that this was something they relied upon so heavily for feedback.

In addition to restricting social interaction, Covid-19 also led to a lower frequency of revisiting skills; in many cases it had only been possible to cover material once during the course.

Support

A consequence of the sense of isolation and not getting to know staff or other students was that some respondents in study 2 found it difficult to ask peers and staff for help. Thirty-seven percent of respondents did not ask anyone for help when completing online activities. Of that group, 43% stated that they didn't ask anyone for help because they did not feel comfortable asking for help and 21% stated that they didn't ask because they did not know how to get in contact with someone.

So it's just … where to go for like help if you need it? … [Y]ou can ask on the discussion board which is good because it's anonymous but then it's waiting for the reply and then you may have to ask another question and the follow up to that. And it's just difficult if you don't really understand what's going on.'

From a staff perspective, the online activities required significantly more input than in-lab sessions, which were predominantly supported within the scheduled session time with support from a technician. For online activities, each weekly cycle required the course coordinator to undertake: the design and writing of the activity and assessment guidance instructions, producing videos/quizzes/assignments, proofreading, collating resourcing, uploading to the VLE, monitoring engagement, training PGs to mark the assignments, returning marks to students and providing feedback. Support for students undertaking the online activities was provided via emails, discussion board posts (outside scheduled session times) along with some synchronous sessions repeated four times per week (P1A) and three times per week (P1B). The number of students commenting on the anonymous discussion board had been less than expected by the course coordinator.

Discussion and Conclusion

The data collected evaluating the implementation of the blended (P1A) and online (P1B) laboratory modules have provided several useful insights for future development of online or blended teaching.

Student Attitudes towards Online Laboratory Teaching

The redeveloped course was carefully designed to retain the learning objectives of the traditional laboratory course whilst separating out the elements that required hands-on experience and those that could be taught effectively outside the laboratory. However, what is clear from the student responses, is that students' did not perceive this to be providing them with an 'authentic' practical chemistry experience. Students' perceptions of what was required in practical work highlighted that they did not consider 'soft' skills (e.g. data use, enquiry based skills) and procedural 'hard' skills to be worthwhile unless they were carried out at the same time as the manipulative and general laboratory 'hard' skills within the laboratory environment. Attempts to manage students' expectations and understanding of the design of the course in the induction session did not stop the students considering the online activities to be less worthwhile than the in-lab sessions. It was, however, interesting to observe that the skills which students reported feeling most able and most confident in by the end of term 1 were predominantly soft and procedural skills, despite these having been taught in the online activities, which they considered to be less worthwhile than the in-lab sessions. The students did note that the range of skills which the online activities had covered had been broad and that the in-lab sessions were not always able to cover the skills which had been targeted in the online activities.

Successful completion of work (as perceived by the students) was reported to be higher for in-lab than online activities. Further evaluation is required to identify whether their perceived completion of tasks matches their assessment scores for the learning objectives for the course.

Flexibility Offered by Online Activities

Students found the flexibility in scheduling their work for the online activities to be beneficial. However this introduced challenges in ensuring all students attended the correct live sessions and required a significant investment of time using multiple modes of communication by the course coordinator to ensure attendance at the correct sessions. Attendance in live online sessions improved through the course, but it was clear some students were confused about the requirements due to the variety in structure each week.

Interactions

A key challenge presented by Covid-19 was the social distancing requirement for students both within and outside the laboratory. Within the lab, it significantly impacted on students' communication with demonstrators due to maintaining 2-metre separation and PPE requirements. Online, students were impacted as they did not feel they knew their peers or staff well enough to feel comfortable asking for help. In the online activities, communication with the demonstrator was significantly less than in the laboratory sessions, with only 2% of students reporting that they communicated with a demonstrator more than three times per activity. Staff were also less likely to initiate contact in the online activities compared to in the lab. Online videos from the demonstrator providing guidance or demonstrating techniques were much more widely engaged with, as was use of the internet for information gathering. The interaction which the highest proportion of students undertook most frequently in both the in-lab and online sessions was reading the written instructions.

Although social distancing is potentially unique to the Covid-19 pandemic for in-lab sessions, there are still important lessons to draw for future online or blended learning courses. Providing opportunities for students to get to know one another and staff at the start of the course (e.g. through ice-breaker activities within induction sessions), facilitating interactions between students (e.g. through group work and Zoom workshops) and staff initiating interactions with students (rather than waiting for students to ask for help) can aid in creating a more effective learning environment, helping students feel less isolated, more connected with their peers and supported by staff. In turn, this may help to improve students' satisfaction with the course, as providing more social support may help students to engage in help-seeking behaviours, which may improve their perceptions of self-efficacy.

There are still some challenges to address for implementation of the Zoom workshops. Firstly, how to encourage participation from all students and to prevent domination of conversation by a small number of individuals. Secondly, how to ensure that students that lack social confidence or that find interaction with strangers challenging do not feel excluded or anxious in facilitated group work sessions. One solution may be to offer an option of a 'quiet' room within the Zoom workshop where people can work alone, with camera and microphone off, but still be able to seek help if needed. It is also important to ensure that the student and staff induction sessions introduce the concept of diversity of personalities and supports how they should aim to be understanding and inclusive in their own actions and behaviors within sessions.

Elements to Retain in Future Courses

Emerging from the pandemic, much of the new approach will be kept. The explicit separation of training in manipulative skills (in the lab) and procedural and soft skills (online) will continue, integrated into coherent activities. Pre- and post-lab will be focused on the soft and procedural skills directly, with related manipulative skills covered in the lab. This could include online independent or group work activities probing the 'thinking behind the doing' related to the manipulative skills. Future laboratory course timetabling might consider including online-only sessions alongside the in-laboratory sessions, to give students structure in their online study and to allow for synchronous teaching sessions to support the online study. These could continue to use the successful Zoom breakout room model.

Adaptions for Future Implementation

Anticipated adaptations to the course content are expected to include more detailed instructions in the induction on the use of the VLE and the skills that the course aims to develop. The perception of the value of the online activities may be clearer to students if the distinction between, and value of, the manipulative, procedural and soft skills is made more explicit in the induction. Expanded training for demonstrators will also emphasise the aims of the course and include the importance and relevance of the online activities as well as the in-lab work.

Integrating ongoing evaluation activities into the course structure will also be carefully considered. Planning in time to periodically collect information from staff and students and to reflect during the course on what is working well, the challenges in implementation, how different groups of students are responding to the course and where adaptions can be made has been extremely beneficial and should not be lost.

Consideration will also be given to the collection of informal feedback about student attitudes and experiences throughout the course. The online activities reduced the opportunity to gauge students' facial expressions and body language. This social interaction between student and teacher is key to developing effective teaching and learning as well as ensuring student wellbeing, therefore it is essential to find a mechanism through which this can be re-established for the online environment.

References

1. Moore, M. G.; Anderson, W. G. *Handbook of Distance Education*; Lawrence Erlbaum Associates: Mahwah, 2003.
2. Singh, V.; Thurman, A. How many ways can we define online learning? A systematic literature review of definitions of online learning (1988-2018). *American Journal of Distance Education* **2019**, *33* (4), 289–306.
3. Smith, K.; Hill, J. Defining the Nature of Blended Learning through Its Depiction in Current Research. *High. Educ. Res. Dev.* **2019**, *38* (2), 383–397.

4. Hofstein, A.; Mamlok-Naaman, R. *The Laboratory in Science Education: The State of the Art*; 2007; Vol. 8.

5. The Quality Assurance Agency for Higher Education. *QAA Membership: Subject Benchmark Statement Chemistry*. https://www.qaa.ac.uk/docs/qaa/subject-benchmark-statements/subject-benchmark-statement-chemistry.pdf?sfvrsn=1af2c881_4 (accessed Jan. 4, 2021).

6. Royal Society of Chemistry. *Accreditation of degree programmes*; Royal Society of Chemistry: London, U.K., 2019.

7. Royal Society of Chemistry. *Letter to Heads of Chemistry 8 September 2020*; Royal Society of Chemistry: London, U.K., 2020.

8. Page, C.; Wilson, M.; Kolb, D. *Managerial competencies and New Zealand managers: On the inside, looking in;* University of Auckland: Auckland, New Zealand, 1993.

9. Turiman, P.; Omar, J.; Daud, A. M.; Osman, K. Fostering the 21st century skills through scientific literacy and science process skills. *Procedia-Social and Behavioral Sciences* **2012**, *59*, 110–116.

10. Weber, M. R.; Crawford, A.; Rivera, D., Jr.; Finley, D. A. Using Delphi panels to assess soft skill competencies in entry level managers. *Journal of Tourism Insights* **2010**, *1* (1), 12.

11. Chadwick, S.; de la Hunty, M.; Baker, A. Developing Awareness of Professional Behaviours and Skills in the First-Year Chemistry Laboratory. *J. Chem. Educ.* **2018**, *95* (6), 947–953.

12. Galloway, K. R.; Malakpa, Z.; Bretz, S. L. Investigating Affective Experiences in the Undergraduate Chemistry Laboratory: Students' Perceptions of Control and Responsibility. *J. Chem. Educ.* **2016**, *93* (2), 227–238.

13. Ausubel, D. P.; Novak, J. D.; Hanesian, H. *Educational Psychology: A Cognitive View*, 2nd ed.; Holt, Rinehart and Winston: New York, 1978.

14. García-Martínez, J. A.; Rosa-Napal, F. C.; Romero-Tabeayo, I.; López-Calvo, S.; Fuentes-Abeledo, E. J. Digital Tools and Personal Learning Environments: An Analysis in Higher Education. *Sustainability* **2020**, *12* (19), 8180.

15. Oskamp, S.; Wesley Schultz, P. *Attitudes and Opinions*; Psychology Press: Sussex, 2014.

16. Marinoni, G.; Van't Land, H.; Jensen, T. The impact of Covid-19 on higher education around the world. *IAU Global Survey Report*; International Association of Universities: Paris, 2020.

17. Bandura, A. *Social Foundations of Thought and Action: A Social Cognitive Theory*; Prentice Hall: Englewood Cliffs, 1986.

18. Bandura, A. Self-Efficacy: Toward a Unifying Theory of Behavioral Change. *Psychol. Rev.* **1977**, *84* (2), 191–215.

19. Berger, C.; Kerner, N.; Lee, Y. *Understanding student perceptions of collaboration, laboratory and inquiry use in introductory chemistry*. http://www-personal.umich.edu/~cberger/narst99folder/narst99.html (accessed Jan. 22, 2021).

20. Compte, O.; Postlewaite, A. Confidence-Enhanced Performance. *Am. Econ. Rev.* **2004**, *94* (5), 1536–1557.

21. Martinez-Jimenez, P.; Pontes-Pedrajas, A.; Polo, J. Learning in Chemistry with Virtual Laboratories. *J. Chem. Educ.* **2003**, *80* (3), 346–352.

22. Chittleborough, G. D.; Mocerino, M.; Treagust, D. F. Achieving Greater Feedback and Flexibility Using Online Pre-Laboratory Exercises. *J. Chem. Educ.* **2007**, *84* (5), 884–888.

23. Chaytor, J. L.; Al Mughalaq, M.; Butler, H. Development and Use of Online Prelaboratory Activities in Organic Chemistry to Improve Students' Laboratory Experience. *J. Chem. Educ.* **2017**, *94* (7), 859–866.
24. Blackburn, R. A. R.; Villa-Marcos, B.; Williams, D. P. Preparing Students for Practical Sessions Using Laboratory Simulation Software. *J. Chem. Educ.* **2019**, *96*, 153–158.
25. Garcia, P. A. *Interaction, Distributed Cognition and Web-based Learning.* https://www.learntechlib.org/p/9371 (accessed Sep. 26, 2019).
26. Hillman, D. C. A.; Willis, D. J.; Gunawardena, C. N. Learner-Interface Interaction in Distance Education: An Extension of Contemporary Models and Strategies for Practitioners. *Am. J. Distance Educ.* **1994**, *8* (2), 30–42.
27. Millar, R. *The Role of Practical Work in the Teaching and Learning of Science*; 2004.

Editors' Biographies

Elizabeth Pearsall

Elizabeth Pearsall (Ph.D., University of North Carolina at Greensboro) is currently the Associate Dean for the Institute for Teaching Excellence at York Technical College, a two-year technical college in Rock Hill, South Carolina, whose mission is to build the community through maximizing student success. Under her leadership, the Institute for Teaching Excellence provides guidance and direction for teaching innovation and excellence throughout the College, for courses in all modalities. Previously, she served as an Associate Professor and the Director of Faculty at a large doctoral granting institution overseeing chemistry, biology, physics, and earth science faculty. In this role, she was afforded an opportunity to train, coach, and evaluate faculty in online teaching excellence principles across multiple science disciplines. With more than 10 years of chemistry teaching experience, she has instructed in multiple modalities including on-ground, hybrid, and online chemistry and biology courses. Rooted in teaching practices developed while instructing chemistry and biology courses, and honed over the years through a combination of experimentation in the classroom, teaching across science disciplines, and additional professional development, her philosophy is that instruction must be relevant, equitable, and focused on student success regardless of discipline or modality.

Kristi Mock

Kristi Mock (Ph.D., University of Toledo) is a chemistry lecturer at the University of Toledo. She started teaching in 2005 during graduate school, and did not expect to fall in love with the art of teaching chemistry. After nine years teaching in a traditional, on-ground chemistry classroom, she entered the world of online teaching. Since 2014, she has been involved in finding engaging, fun, and interactive methods of promoting student learning in the online chemistry classroom.

Matthew Morgan

Matt Morgan (Ph.D., Montana State University) currently instructs chemistry courses in fully online environments at Western Governors University and Embry Riddle Aeronautical University. Prior to teaching fully online, he instructed hybrid chemistry courses starting in 2011 where he developed a keen sense of the skills needed to successfully instruct chemistry students from afar. After graduating from the Air Force Academy, he went on to serve his country for 22 years in the Air Force. As a KC-135 aircraft commander, he flew during the Cold War and Desert Storm. Previously, he was a Professor of Chemistry at Hamline University and the United States Air Force Academy for 21 years.

Brenna A. Tucker

Brenna Tucker (Ph.D., University of Alabama at Birmingham) currently serves as the Introductory Chemistry Coordinator at the University of Alabama at Birmingham. She has developed multiple lecture, laboratory, and recitation chemistry courses for online instruction. Although she always knew she wanted to be an educator, she discovered a passion for chemistry education as a teaching assistant in graduate school. Her postdoctoral pursuits focused on developing an online chemistry curriculum and she has completed professional development in the areas of online pedagogy and teaching practices.

© 2021 American Chemical Society

Indexes

Author Index

Benore, M., 45
Casselman, M., 21
Chiu, B., 123
Clulow, A., 105
Connor, M., 177
Corkish, T., 77
Cramman, H., 177
Curtin, A., 135
Davidson, M., 77
Erdmann, M., 59
Faulconer, E., 165
Genova, L., 1
Haakansson, C., 77
Hau, C., 177
Jenkinson, S., 93
Kiappes, J., 93
Kloepper, M., 1

Kloepper, K., 1
Lapeyrouse, N., 123
Lopez, R., 77
Marincean, S., 45
Mock, K., x
Morgan, M., x, 165
Pearsall, E., x, 59
Ratnayaka, S., 59
Robson, J., 177
Sarju, J., 135
Schatzberg, W., 35
Short, J., 105
Simon, L., 1
Spagnoli, D., 77
Tucker, B., xi, 59
Watson, P., 77
Yuriev, E., 105

Subject Index

E

Emergency remote teaching, student experiences and perceptions
- conclusion, 131
  - emergency remote classes, transition, 132
- introduction, 123
- literature review, 124
  - non-academic concerns, 125
- method, 126
  - codebook, 127t
- results and discussion, 127
  - less accessible, online transition classes, 130f
  - online transition classes, 130f
  - previous online courses, 128f
  - spring 2020, online experience, 131f
  - STEM courses, transition, 129f

F

First semester general chemistry flipped remote classroom
- conclusion, 43
- discussion, 41
- introduction, 35
- methods, 37
- results, 38
  - average student first semester general chemistry grades, 39t
  - metacognition response rates, 39t
  - sample metacognition reflection questions, 39t
- study limitations, 40

H

High-stakes to low-stakes assessment for online courses, transitioning, 21
- frequent online homework, 27
  - student impressions, 27
- introduction, 21
- lessons learned towards future instruction, applying, 31
  - demonstrating mastery, purposes, 32
  - low-stakes online quizzing and polling, 24
    - traditional synchronous lectures online, 25
  - online classroom, active learning, 26
- remote instruction, modifying assessment, 22
  - course overview, 23t
  - traditional and remote learning assessments, comparison, 24t
- rethinking online exams, 29
  - ranking-based multiple-choice problem, 31f

I

Implementing blended and online undergraduate chemistry laboratory teaching during the Covid-19 pandemic, lessons learned
- discussion and conclusion, 189
  - future implementation, adaptions, 191
  - interactions, 190
- Durham University, redeveloped first-year undergraduate chemistry laboratory course, 181
  - P1A and P1B modules, structure, 182f
  - P1A in-lab, materials, resources and support arrangements, 183f
- introduction, 177
  - chemistry practical work competences, subdivision, 179t
  - laboratory delivery, definitions for the modes, 178t
  - student interactions in a chemistry laboratory, summary, 180f
- method, 183
  - 2020/21 first year undergraduate chemistry cohort, research studies, 184t
- results, 184
  - challenges and barriers, 188
  - choosing ratings from 1 to 5, perceptions, percentage of respondents, 186f
  - complete in-lab and online sessions, time taken, 187t
  - interactions students carried out more than three times per in-lab or online session, greatest percentage, 185t
  - respondents to survey 2, skills, 187t

O

Online chemistry courses-lessons learned, maintaining rigor
  closing statements, 72
    three different institutions, summary of rigor, 73t
  institution backgrounds, 61
  introduction, 59
  rigor, evolution, 69
  rigor in assessment, 65
  rigor in course design, 62
    York Technical College, 62
  rigor in lab, 67
Online chemistry laboratory instruction, options and experiences
  conclusions, 174
  distance wet chemistry options, 166
    commercial home kits, 169
    distance wet chemistry approaches, comparison of features, 167t
    online dry lab options, 170
    online laboratory modalities, comparison of features, 171t
    students, passive approach, 168
  introduction, 165
  laboratory choices comparison, 173
  laboratory instruction options, 166
Online environment, adapting group-based problem solving
  future, lessons learned and ideas, 99
    acutal and perceived learning, 102
    DNA Alkylation Workshop, student Miro board, 101f
  introduction, 93
    DNA alkylation, 96
    DNA alkylation agents, 97f
    on-site workshops, 94
    student-interaction, 95
    sugars and mimetics, 97f
  pivot to online, 98
Online format, problem based learning group projects
  background, 45
  conclusions, 55
  online group learning, our approach, 47
  projects, 48
    nonmajors biochemistry, upper division course, 53

organic chemistry, sophomore course, 50
Online pre-laboratory lightboard videos, how-to guide
  conclusion, 88
  equipment and setup, 83
    budget, 85
    equipment and set up for filming lightboard videos, image, 84f
    setup of a lightboard recording studio, approximate costs, 86t
  filming, tips, 86
  introduction, 77
    lightboard video, screenshot, 78f
  student responses, 87
  video design, 79
    information being presented on a single lightboard, contrast, 82f
    lightboard before and after working through an error analysis example, image, 81f
    talking head introductory portion of the video, screenshot, 80f

P

Promoting student learning and engagement
  conclusions and future work, 15
  course content, student interactions, 6
    balancing workload, 7
    sample nonmajors chemistry lab learning outcomes, 9f
    student learning, online assessments, 8
  in-person and hybrid classes, application, 13
    group role descriptions and examples of interactions, 14f
  introduction, 1
    common concerns about online courses, 2t
  online laboratory courses, additional considerations, 13
  student-instructor interactions, 3
    midterm check-in, 5
    midterm check-in, example, 6f
    online courses, guide for communication plan development, 4t
    student introduction assignment via a discussion board, example, 4f
  student-peer interactions, 10
  what do students want in an online course, 11

S

Students as partners, 135
- conclusion, 154
- engagement and experience in learning and teaching, impact of partnership, 146
  - relationship building, 147
- equality, diversity, inclusion, and respect, spotlight, 148
  - examples from practice, 150
  - student partnerships, 149
- future practice, recommendations, 151
  - balance of power, addressing, 152
  - co-creation in higher education, examples of key good practices, 151f
  - employ inclusive recruitment and working practices, 153
- introduction, 136
  - author positionality, 138
  - higher education online learning, design and implementation, 137
  - student roles in education design, onion model, 136f
- online learning and teaching in chemistry, collaborative design and creation
  - co-creative teams, team working and collaboration, 144
  - examples of student participation in curriculum and learning design, pictorial representations, 140f
  - pedagogical partnerships, 141
  - successful student-instructor partnerships, definitions of the five criteria, 142t
  - whole-cohort approaches, 143

U

Upside down world, teaching chemistry down, 105
- context and data collection, 106
- dilemmas, decisions, and teaching implications, 107
  - discovery, animations, 108f
  - interactive lecture components, snapshot, 109f
  - interactive lectures, possible delivery modes considered, 108f
  - keep-start-stop questions, student and TA responses, 113t
  - one of the video demonstrations, front-on screen shot, 111f
  - open-ended survey questions, common themes in student responses, 114t
  - SCORM exercise on acids and bases, snapshots, 112f
  - student engagement with recorded lectures, 110f
  - survey questions at the end of the academic year, summary of student responses, 114f
  - TA responses to interview questions, common themes, 116t
  - weekly activity table, example, 112f
- stakeholder perspectives, 118
- summary, 120

Printed in the USA/Agawam, MA
August 9, 2022

796900.001